Socioeconomic, Sustainability, and Human Factors in Transit 2018

Transportation Research Record: Journal of the Transportation Research Board Volume 2672, Issue 6

ISBN: 978-1-5108-8176-1

Printed from e-media with permission by:

Curran Associates, Inc.
57 Morehouse Lane
Red Hook, NY 12571

proceedings.com

Some format issues inherent in the e-media version may also appear in this print version.

Published by Sage Publications USA

ISBN (Print) 978-1-5108-8176-1
ISBN 2018 Printed Set (All Issues) 978-1-5108-7735-1

Printed by Curran Associates, Inc. (2019)

Additional copies of this publication are available from:

Curran Associates, Inc.
57 Morehouse Lane
Red Hook, NY 12571 USA
Phone: 845-758-0400
Fax: 845-758-2633
Email: curran@proceedings.com
Web: www.proceedings.com

Volume 2672 Number 6 December 2018

TRR
JOURNAL OF THE TRANSPORTATION RESEARCH BOARD

Contents

Socioeconomic, Sustainability, and Human Factors in Transit 2018

Articles

---

* Membership as of February 2019.

TRB
TRANSPORTATION RESEARCH BOARD

CORE SPONSORS

Transportation Departments of the 50 States and the District of Columbia

Federal Government

U.S. Department of Transportation
- Federal Aviation Administration (FAA)
- Federal Highway Administration
- Federal Motor Carrier Safety Administration (FMCSA)
- Federal Railroad Administration (FRA)
- Federal Transit Administration
- National Highway Traffic Safety Administration (NHTSA)
- Office of the Assistant Secretary of Transportation for Research and Technology (OST-R)

U.S. Air Force Civil Engineer Center (AFCEC)
U.S. Coast Guard (USCG)
U.S. Department of Energy
U.S. Department of the Interior (DOI)
U.S. Environmental Protection Agency (EPA)

Nongovernmental Organizations

American Public Transportation Association (APTA)
Association of American Railroads (AAR)
California Air Resources Board

The National Academies of

SCIENCES · ENGINEERING · MEDICINE

The **National Academy of Sciences** was established in 1863 by an Act of Congress, signed by President Lincoln, as a private, nongovernmental institution to advise the nation on issues related to science and technology. Members are elected by their peers for outstanding contributions to research. Dr. Marcia McNutt is president.

The **National Academy of Engineering** was established in 1964 under the charter of the National Academy of Sciences to bring the practices of engineering to advising the nation. Members are elected by their peers for extraordinary contributions to engineering. Dr. C. D. Mote, Jr., is president.

The **National Academy of Medicine** (formerly the Institute of Medicine) was established in 1970 under the charter of the National Academy of Sciences to advise the nation on medical and health issues. Members are elected by their peers for distinguished contributions to medicine and health. Dr. Victor J. Dzau is president.

The three Academies work together as the National Academies of Sciences, Engineering, and Medicine to provide independent, objective analysis and advice to the nation and conduct other activities to solve complex problems and inform public policy decisions. The Academies also encourage education and research, recognize outstanding contributions to knowledge, and increase public understanding in matters of science, engineering, and medicine.

Learn more about the National Academies of Sciences, Engineering, and Medicine at **www.national-academies.org.**

The **Transportation Research Board** is one of seven major programs of the National Academies of Sciences, Engineering, and Medicine. The mission of the Transportation Research Board is to increase the benefits that transportation contributes to society by providing leadership in transportation innovation and progress through research and information exchange, conducted within a setting that is objective, interdisciplinary, and multimodal. The Board's varied activities annually engage about 7,000 engineers, scientists, and other transportation researchers and practitioners from the public and private sectors and academia, all of whom contribute their expertise in the public interest. The program is supported by state transportation departments, federal agencies including the component administrations of the U.S. Department of Transportation, and other organizations and individuals interested in the development of transportation.

Learn more about the Transportation Research Board at **www.TRB.org**.

Article

TRR
JOURNAL OF THE TRANSPORTATION RESEARCH BOARD

Transportation Research Record
2018, Vol. 2672(6) 1–10

DOI: 10.1177/0361198118790327
journals.sagepub.com/home/trr
SAGE

Reducing Rural Car Ownership: Cultural Not Policy Changes?

Emma L. Rearick[1] and Gregory L. Newmark[1]

Abstract

Automobile use is recognized as affecting public health, environmental sustainability, land use, and household expense. Car use is closely tied to car ownership rates. Most car ownership research focuses on urban areas; however, 97% of the United States' land area and a fifth of its population remains rural. Factors that affect car ownership in these communities may be different than in more urbanized areas. This research focuses on the 2,285 counties in the continental United States that are defined as entirely rural by the guidelines established in the Agricultural Act of 2014. These counties were grouped by five multi-state regions using U.S. Census Bureau definitions. Their percentage changes in car ownership, as well as other demographic variables, over a quarter century were calculated using data from the 1990 Decennial Census and the 2014 5-Year American Community Survey. A multiple regression model was estimated for each grouping to identify counties with lower-than-expected changes in car ownership. For each grouping, one of these outlying counties was selected and matched with another county whose changes in car ownership were within expected ranges given demographic developments. Local professionals were then interviewed to identify policies possibly responsible for the difference in car ownership trends between the matched-pair counties. The interviews suggested that, contrary to expectation, transportation policies had no discernable effect on rural car ownership, but land use polices and, more often, cultural factors linked to changing populations were associated with reduced rural car ownership.

Automobile use is associated with a host of public health, environmental sustainability, local land use, and household economic externalities. Since car ownership is strongly correlated with car use (*1*), transportation planners and researchers often study vehicle ownership to glean lessons on how to reduce both ownership (*2*, *3*) and use. Such studies typically focus on urban areas, home to the most extensive transit and non-motorized networks that can provide alternatives to driving; however, 97% of the U.S. land area (*4*) and almost 20% of the U.S. population is rural (*5*) – areas and people that are almost entirely ignored in most sustainable transportation research.

This research seeks to undo this omission by exploring the factors associated with rural reductions in car ownership.

Literature Review

Need for Rural Research on Car Ownership

Most car ownership studies focus on households in urban areas (*6–8*). Rural areas are often overlooked by transportation researchers, perhaps because of the misconception that there are no alternatives to driving in rural areas. The limited relevant studies consider rural areas within larger data sets (*9*, *10*), explicitly compare rural and urban areas (*11*, *12*), or focus primarily on rural areas (*13*).

Some argue that rural travel is sufficiently distinct from urban travel to warrant its own research and policies (*13*, *14*). For example, since households in rural areas of the United Kingdom are relatively insensitive to fuel or car purchase cost increases, they are disproportionately affected by policies, such as surtaxes, that raise these costs. Explicit consideration of rural households might lead to more equitable policies for all, such as congestion pricing (*12*). Furthermore, policies effective in shifting modes away from driving in urban areas, such as raising car ownership costs to encourage transit use, penalize rural residents lacking other transportation options without achieving the actual policy goal (*13*).

[1]Kansas State University, Manhattan, KS

Corresponding Author:
Address correspondence to Gregory L. Newmark: gnewmark@ksu.edu

Factors Associated with Car Ownership

Regardless of the urban or rural context, car ownership is typically studied at the household level. Pyddoke and Creutzer noted that individuals often have access to household cars that may not be registered in their name, so studying individuals alone would not reflect whether they contribute to purchasing and using a car (*11*). Income, too, is usually shared among individuals in a household, and income is strongly correlated with car ownership (*12*, *15–17*) for both rural and urban households (*11*).

Age of the head of household and car ownership are also thought to be correlated. Pyddoke and Creutzer found that car ownership is negatively affected if at least one adult in a household is older than 66 years or younger than 25 (*11*). Schimek agreed that households with young heads had lower rates of car ownership, but defined under 35 as young (*17*). Dargay noted that over a lifetime, car ownership is fairly low for younger adults, and increases rapidly as they age (*12*). Once any children begin to move out of the house and the head of the household begins to approach retirement, car ownership rates begin to drop again. But this trend may not be attributable simply to age, because age is also correlated with income and family composition. Oakil et al. closely examined life cycle events and car ownership to concur with Dargay that as a head of household ages, households tend to increase in size due to marriage and births, events which lead to increased car ownership (*12*, *18*).

Population density is often correlated with car ownership. Although, there have been mixed conclusions about how strongly density by itself is correlated with car ownership, perhaps because density is often a proxy for other factors that discourage car ownership (*6*, *17*, *19*, *20*). Such factors include parking availability, proximity to transit, and congestion. Sun et al. determined that the built and economic environment has the most influence over household car ownership, followed by the time period and age of household members (*10*).

Car ownership rates are not static. Certain generations may be more inclined to own cars, although as a cohort ages those trends may change (*10*, *21*). Many studies use longitudinal data to study and predict change in car ownership levels over time on both the national and household level (*12*, *22*).

Alternative Experimental Designs

Most of the research on car ownership and use relies on statistical models, which although powerful are not always well suited to real-world settings. An alternative approach to explore the effects of local policies on driving has been the quasi-experimental matched-pair approach. Matched pairs have been used to compare car ownership and use in two different neighborhoods that share many demographic and structural elements but not the variable to be tested (*23*, *24*). This approach does not yield a perfect control. Cervero and Gorham provide some guidance for picking pairs; for example, the median household incomes of the paired regions should be within 10% of each other (*23*).

Despite a lack of study of car reduction policies for rural areas, there has been a renewed interest in rural transportation planning. The Federal Highway Administration's publication *Planning for Transportation in Rural Areas* encourages a multimodal approach and notes that "bicycling and walking are often the 'forgotten modes' of transportation planning—especially in rural areas" (*25*). A more recent publication by the Federal Highway Administration, *Small Town and Rural Multimodal Networks* directly addresses the challenges of designing multimodal infrastructure for rural areas, but provides few policy guidelines (*26*).

This work seeks to strengthen the research basis for rural car reduction policies by focusing explicitly on rural areas and using an innovative modeling structure to control for demographic factors to generate a matched-pair comparison.

Methodology

This study combines quantitative and qualitative approaches in two sequential stages.

In the first stage, a multiple regression analysis identifies counties from five regions, based on the established U.S. census regions and divisions, whose car ownership rate has either dropped much more or grown much less than expected while controlling for socio-demographic changes between 1990 and 2014. This approach posits that the unexplained difference might be due to local policies. These outlier counties are then paired with a control county in the same state whose car ownership rate changes are in line with expected values from the statistical model.

In the second stage, planning officials in the 10 counties are contacted to identify any policies over the last quarter century that may have contributed to the distinction in outcome between the outlier and control counties.

Modeling Changes in Rural Car Ownership Rates

The Agricultural Act of 2014, commonly known as the Farm Bill, defines rural areas as those with a population up to 35,000 and rural in character (*27*). This study identified counties and county equivalents for which all land area meets that Farm Bill definition as rural. Counties and county equivalents were chosen as the unit of analysis because the boundaries of these jurisdictions are very stable – facilitating longitudinal studies – and the data on car ownership are available. A total of 2,285 out of the 3,108 counties in the continental United States met this rural criterion.

Demographic data from the 1990 Decennial Census and the 2014 5-Year American Community Survey, the most recently available at the time of the study, were gathered (*28*, *29*). The year 1990 was chosen as the base year because 24 years is long enough to establish ongoing trends, but still recent enough that cultural factors, such as two-earner households, would remain relatively comparable. Data on a wide array of variables typically correlated with car ownership rates were retrieved at the county level, including age breakdowns, income, employment, family size, and presence of children.

A linear regression was estimated for each of five regions of the United States: the Midwest, Northeast, Southeast, West, and West South Central. The analysis was done by region to account for factors such as settlement patterns that are shared among regions. These multi-state regions were chosen by first estimating models for counties in each of the established nine U.S. Census Bureau divisions. Neighboring divisions with similar intercepts were then grouped into a final set of five study regions shown in Figure 1, below.

Modeling and selection by region ensures that not all counties of interest are grouped into one state or area of the country. The dependent variable was the percent change in household car ownership rates, by county, between 1990 and 2014. These rates are reported by the U.S. Census Bureau in aggregate for an entire county. They do not independently consider household size or composition. Figure 2 shows that the majority of counties (gray bars) saw an increase in their household car ownership rate during this time, although some did decrease (purple bars).

The independent variables were also all measures of percentage change over that same period, including median household income (in nominal terms), total population from 1990 to 2014, share of the population aged 65 and older, and share of the population aged 17 and younger. The models are presented in Table 1. These measures are meant to account for socio-demographic changes (although total population is also a rough proxy for land use density) to attribute the remaining explanatory power to possible policy interventions.

All the models are significant with the models accounting for between a sixth and a third of the variance.

In all regions, rising incomes were associated with rising car ownership rates although the relationship was substantially stronger in the Northeast. In two regions, the Northeast and Midwest, growth in the population share of senior citizens was associated with a reduction in car ownership rates whereas in one region, the Southeast, growth in the population share of children was associated with a reduction in car ownership rates. In one region,

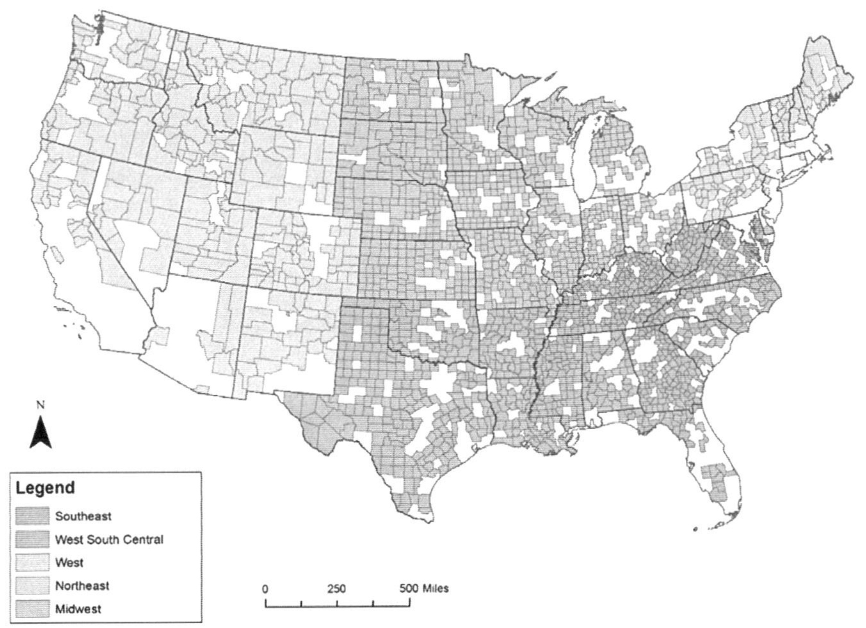

Figure 1. Rural counties by study region.

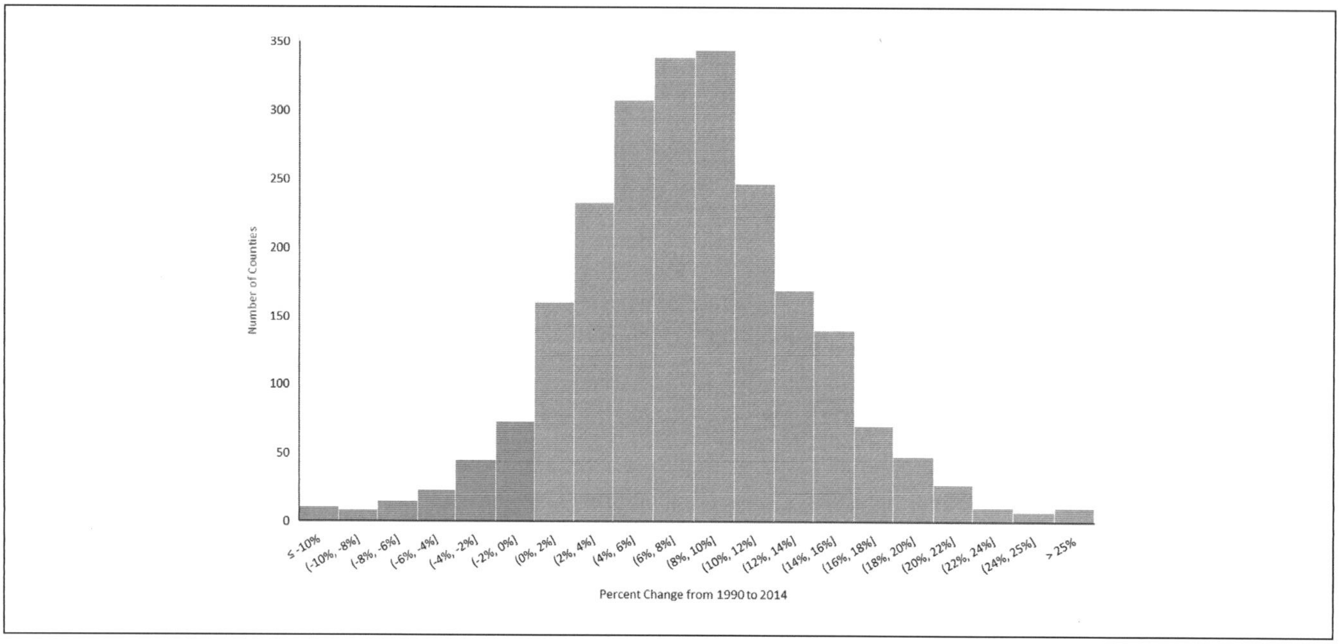

Figure 2. Percent change in rural county car ownership rates from 1990 to 2014.

Table 1. Models of Rural County Car Ownership Change by Study Region

Variables	Midwest		Northeast		Southeast		West		West South Central	
%Δ from 1990 to 2014	Est (SE)	Sig	Est (SE)	Sig	Est (SE)	Sig	Est (SE)	Sig	Est (SE)	Sig
(Intercept)	−1.725	*	−6.843	*	−2.987	**	−5.163	**	−6.247	***
	(0.785)		(3.055)		(0.907)		(1.685)		(1.685)	
Share of seniors (65 +)	−0.044	***	−0.053	*						
	(0.010)		(0.027)							
Share of children (0–17)					−0.075	**				
					(0.024)					
Total population									0.047	***
									(0.013)	
Median income	0.400	***	0.778	***	0.499	***	0.517	***	0.545	***
	(0.031)		(0.142)		(0.042)		(0.076)		(0.066)	
Summary statistics										
Adjusted R^2	0.215		0.316		0.211		0.171		0.209	
F-statistic	113.8		24.5		87.2		46.07		41.17	
Degrees of freedom	(2, 821)		(2, 100)		(2, 641)		(1, 217)		(2, 302)	
p-Score	0.000		0.000		0.000		0.000		0.000	

Note: The dependent variable is percent change in county car ownership. Some counties fell out of analysis due to missing variables. Est = estimated coefficient; SE = standard error.

the West South Central, growth in total population was associated with growth in car ownership rates.

Matched-Pair Selection

Counties of interest were identified as those with standardized residuals more than 2.0 standard deviations below the mean, indicating that car ownership rates in the county had decreased faster (or increased slower) than would be expected according to the regional models. Comparison counties to serve as controls were identified as those with standardized residuals between −0.2 and 0.2, indicating that the car ownership rates in the counties had performed as the regional model predicted. Figure 3 presents this idea graphically.

A list of counties of interest was generated for each of the five study regions, ranging from a low of three counties for the Northeast region to a high of 29 counties for

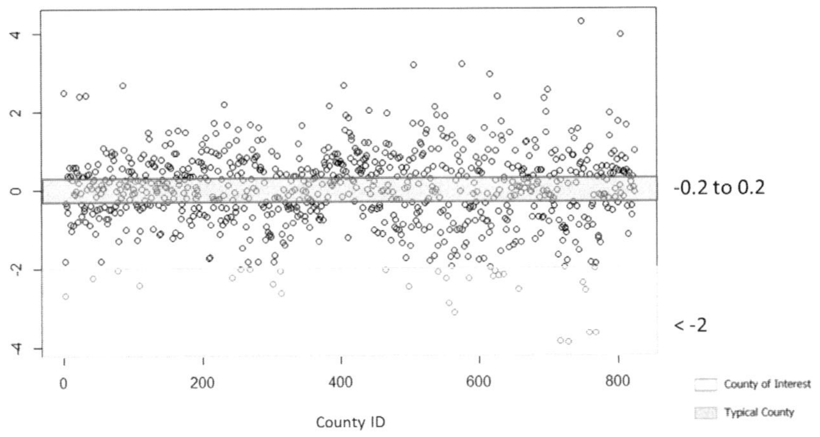

Figure 3. Illustration of process used to identify cases (Midwest region).

Figure 4. Selected matched-pair counties for qualitative analysis.

the Midwest region. A list of comparison counties was also generated, ranging from 22 counties for the Northeast region to 131 for the Midwest. All counties with a total population of less than 10,000 in both 1990 and 2014 were eliminated from the lists, except in the West region where none of the counties of interest met the minimum population threshold.

One county of interest was then selected for each region, with careful attention to avoid counties with unique or very uncommon characteristics, such as an isolated island location. A comparison county was then selected from the same state, to minimize variation in state-level climates or policies, and with comparable income and population levels in both 1990 and 2014. Figure 4, below, presents the five matched pairs. Comparative information on each pair is presented in Table 2, below.

Comparing Rates

It is important to note that although the identified counties are outliers, their actual car ownership rates and

Table 2. Comparative Information on Matched-Pair Counties

Study region	Midwest		Northeast		Southeast		West		West South Central	
Selected state	Indiana		New York		Georgia		Montana		Texas	
County type	Interest	Typical	Interest	Typical	Interest	Typical	Interest	Typical	Interest	Typical
County name	Adams	Huntington	Yates	Greene	Union	Gilmer	Chouteau	Custer	Eastland	Terry
Car ownership change										
1990 cars per household	1.902	1.852	1.678	1.712	1.964	1.820	2.312	1.871	1.775	1.741
2014 cars per household	1.807	1.919	1.703	1.809	1.871	1.970	2.056	2.021	1.654	1.794
Absolute change	−0.095	0.067	0.025	0.097	−0.092	0.151	−0.256	0.150	−0.121	0.053
Percent change	−4.984	3.592	1.471	5.659	−4.700	8.277	−11.075	8.036	−6.827	3.064
Std. residual from model	−2.032	0.019	−2.236	−0.152	−2.633	0.013	−2.865	0.031	−2.782	−0.075
Demographic data										
1990 population	31,095	35,427	22,810	44,739	11,993	13,368	5,452	11,697	18,488	13,218
2014 population	34,533	36,959	25,281	48,618	21,553	28,441	5,859	11,869	18,403	12,681
1990 median household income (US$)	28,790	29,680	24,870	27,470	20,280	21,410	22,360	21,350	15,770	22,390
2014 median household income (US$)	47,964	47,356	50,061	49,864	39,179	39,581	41,270	46,125	35,221	39,494

Note: All data from U.S. Census Bureau; 1990 data is from 1990 Decennial Census; 2014 data from 2014 5-Year American Community Survey (28, 29).

related changes are often not so different from typical counties. As shown in Table 2, the car ownership rates for each of the 10 counties hovers around two cars per household.

The two counties in the West region have remarkably similar 2014 car ownership rates at 2.056 and 2.021, although the first has decreased since 1990 and the second has increased. Regardless of whether a rate increased or decreased, most of the selected counties saw a change of approximately a tenth of a car per household.

Qualitative Comparison

One county planning department employee was contacted via phone or email in each of the 10 counties, where available. If no county planning department was publicly listed, or the department was unresponsive, an employee of another local government or organization was contacted. The employee was asked to recommend someone knowledgeable about local transportation policies and projects. Based on these recommendations, an appropriate interviewee was contacted in each county and asked a series of 14 open-ended questions regarding transportation trends or other changes in the county. Some interviewees shared key documents, including recent survey results, comprehensive plans, or transportation plans. The interviews and planning documents were used to pinpoint any local policies or trends that might affect car ownership rates.

It was expected that clear policies would be identified in the counties of interest that would help explain their status as statistical outliers regarding car ownership rates; however, in only two cases was such a proverbial "silver bullet" found. Instead, and surprisingly, shifting cultural practices linked to changing demographics often appeared to be the critical driver. Most notably, many of the outlier regions saw growth in agricultural religious communities, such as the Amish, whose cultural practices substantially curtail motorized vehicle ownership.

All of the counties examined provide school buses for school children living outside the town in which schools were located, and many also provided busing for students inside town boundaries. All of the counties examined provided paratransit services for older adults and disabled residents, and many opened those services to all residents. For the sake of brevity, only variations on these norms are discussed in the cases below.

Midwest Region

Adams and Huntington counties in Indiana were chosen as the matched pair for the Midwest region. Transportation policies were similar. Both counties recently imposed wheel taxes. Unlike Adams County,

Huntington County, the control county, opens its paratransit to everyone. Land use policies, however, vary.

Adams County has used Community Development Block Grant funds to pay for streetscape improvements in downtown Decatur, the main county population center. Decatur itself sets aside US$50,000 every year for a Sidewalk Assistance Program, which provides a 50/50 match on residential sidewalk improvements. The program pays for half the cost of the sidewalk and the full cost of demolition (*30*). These policies may facilitate reduced car ownership. By contrast, although Huntington County has a 1,200-student university and some concentration of mixed-use buildings in the urban core (as well as a mixed-use zoning district), the city has grown substantially and allows developers to waive sidewalk construction requirements.

There may also be cultural differences contributing to the different car ownership trends. Adams County has a significant Amish presence. Local estimates are around 12,000, although a 2010 estimate from the Association of Religion Data Archives lists 6,343 adherents (*31*). Amish communities resist modern technological conveniences and are less likely to own motorized vehicles although they do travel public roads via horse-and-buggy. (Adams County passed its wheel tax, in part, to raise revenue from Amish who do not contribute via motor fuel taxes to the county roads they use.)

Northeast Region

Yates and Greene counties in New York were chosen as the matched pair for the Northeast region. Both counties recently enhanced local transit offerings. In Yates County, members of the general public were allowed to use the existing paratransit system on January 1, 2017 (*32*). Although no intercity bus providers service the area, the Yates Transit Service does connect to public transportation routes in adjacent counties. In Greene County, public transportation service was substantially expanded beyond paratransit in June of 2016 to include seven fixed routes (*33*). Greene County also offers intercity service via Adirondack Trailways (*34*). Both counties provide school buses, although residents of Penn Yan, Yates County's central village, must walk to school. These transit developments are too new to be reflected in the change in car ownership.

However, once again, cultural differences stand out. Yates County has a population of Old Order Mennonites (*35*), who, while more willing to own cars than the Amish, still report very low rates of car ownership and very high rates of horse-and-buggy ownership. Greene County has no similar concentration of religious adherents who eschew automobile ownership.

Southeast Region

Union and Gilmer Counties in Georgia were chosen as the matched pair for the Southeast region. Both counties report very high population growth rates over the study period and have park-and-ride lots to support commuting to Atlanta. The main distinction appears to be that Union County has become a popular retirement destination. Furthermore, the development in Union County appears to be concentrated in the central city of Blairsville due to geographic and land ownership constraints facilitating shorter trips and alternatives to driving.

West Region

Chouteau and Custer counties in Montana were chosen as the matched pair for the West region. Choteau County is in north central Montana, while Custer County is in the eastern part of the state. The population of Custer County is approximately twice that of Chouteau County, and its median household income is higher.

Transportation policies are similar, although Custer County offers intercity bus service through Greyhound. The main differences affecting car ownership seem to be due to land use policy and culture. Chouteau County is located along the Missouri River and has a strong tourist appeal. To manage the negative impacts of tourism, the county adopted a strong growth management policy in 1985 (*36*). In addition to strong growth management, the county is attractive to Hutterites, who have purchased substantial acreage. The Association of Religion Data Archives notes 100 adherents in 2000, an increase from no adherents in 1990 (*37*) suggesting further increases through 2014, the end of the study period. Hutterites own property communally (*38*) with the results that vehicles are shared widely and car ownership rates decline. Neither strong growth management nor a cultural change has occurred in Custer County.

West South Central Region

Eastland and Terry counties in Texas were chosen as the matched pair for the West South Central region. Eastland County is in the north central part of the state whereas Terry County is located toward the western edge. The population of both counties decreased between 1990 and 2014, but only slightly. The median household income in Eastland County more than doubled, whereas income in Terry County increased more modestly.

There were no clear reasons for the different trends in car ownership between the two counties that emerged from the interviews. Eastland County has seen some small bicycle and pedestrian improvements to improve access to schools and implemented zoning in 2007 but without allowances for mixed-use development. School bus service is available for students who live outside of the city limits. Students with disabilities within the city limits may also ride the bus. Terry County has seen job growth, but most well-off workers prefer to live outside the county and commute. Despite the availability of school buses, many families drive their children to school locally or to schools in surrounding communities. Once again, none of these differences seems to account for the substantial difference in car ownership trends.

Discussion

Land Use Policies

Although no transportation policies stood out as central to reducing car ownership, two counties reported land use changes that were conducive to concentrating growth and likely to be for reducing the need for driving: Union County in the Southeast region and Chouteau County in the West region. Though Chouteau County's policy was explicitly aimed at managing growth, Union County had natural geographic and land ownership constraints that produced a similar outcome.

Cultural Factors

The goal of the modeling step was to select counties of interest that were not demographically exceptional, in order to highlight policies that may impact household car ownership rates. However, of the five counties of interest, four contain populations that tend to own cars at lower rates. Adams County, Indiana is home to a significant Amish population; Yates County, New York contains an Old Order Mennonite population; Union County, Georgia is home to a disproportionate number of older residents; and Chouteau County, Montana has seen Hutterites establish colonies since 1990. Eastland County, Texas is the one county of interest that does not appear to be demographically exceptional. In rural counties with low populations, even a modest cultural change can have a significant impact. The typical counties of each matched pair do not show similar demographic characteristics, which indicates that the exceptional demographic factors could explain the unexpected changes in car ownership rates.

Units of Analysis

As stated earlier, counties are a convenient, stable, and relevant geographic unit. Their borders remain consistent over time, and they correlate with political jurisdictions, unlike census tracts, which may change with each census and are not related to any government unit, or cities that may annex land and residents. However, each county contains multiple neighborhoods, communities, and governments that complicate case studies. For this study, some professionals were familiar with the entire county, but perhaps not trends within incorporated communities, whereas others were familiar with a particular incorporated community but perhaps not the county as a whole. Especially because many policies are implemented within a city or town, it may have been easier to tease out the impacts of a particular program or policy if the case studies had focused on units smaller than counties. Using a smaller geographic unit may be feasible for subsections of the country, or in a study focused on 1 year and not change over time.

Study Limitations

The definition of rural used to eliminate counties from the dataset is potentially too limiting and also too generous. Too limiting in that there are many small communities that are rural in nature that happen to be in a county that also contains an area defined as urban. Because counties were the chosen unit, these areas were removed from the dataset along with the urban area. The selection definition was perhaps too generous in that a small community, rural in nature, could have been included in the dataset while also being very close, geographically and economically, to an area defined as urban in the adjacent county.

For obvious reasons, this is not a double-blind study. The counties were selected and locals interviewed with the knowledge that some would serve as controls and others would serve as test cases. Interviewees were all given and asked the same information, but the interviewer knew if a county was categorized as a control or an outlier. Therefore, it was a challenge as a researcher not to automatically and potentially incorrectly identify policies as correlated with higher or lower car ownership rates.

Even though counties with fewer than 10,000 residents were generally eliminated from the study set, it was still difficult to reliably reach someone knowledgeable about transportation trends and policies at either the county or city level for some counties. Many practitioners were surprised that someone from out of state was interested in their community.

The models were designed to capture change in car ownership rates over time, in order to find communities that may be less car-dependent now than there were in the past. However, interviews conducted in the present day may not effectively capture community changes over time, either because an individual is not familiar with the history, or current events tend to overshadow long-term changes.

Conclusion

The goal of this study was to combine quantitative and qualitative methods to discover and document rural communities that had policy lessons to share about reducing dependence on personal car ownership. Surprisingly, these reductions appeared to be largely linked, not to transportation policies, but to cultural factors tied to specific demographic changes – most commonly growing populations of religious agricultural communities like the Amish, whose beliefs prohibit or limit the ownership of motorized vehicles. The cultural practices of such growing populations can have a substantial impact on car ownership trends in rural counties with typically small populations. Land use practices, both explicit and natural, to concentrate growth also appeared to influence rural car ownership trends – a finding with important implications for policy. Cultural and land use practices may also combine effectively, as occurred with explicit growth management and the Hutterite population in Chouteau County, Montana and with naturally constrained growth and the retiree population in Union County, Georgia. More research needs to explore these opportunities.

Future Research

This study offers several directions for future related research. Fundamentally, it identifies the importance of culture in affecting rural car ownership. Exploring the travel behaviors and built environments of specific low-car ownership populations may be useful for gleaning insights about effectively living car-free or car-lite in rural communities.

This study also presents a rigorous quantitative method for case selection as a basis for subsequent qualitative exploration. This method may be enhanced by focusing on smaller regions of the country or smaller units of analysis, such as towns, to reduce geographic variability; by including demographic variables more sensitive to changing culture, such as religious adherents; by considering employment variables such as proportion of residents working in agriculture or manufacturing; by increasing the vetting criteria for selecting cases; and perhaps by comparing clusters rather than pairs.

Acknowledgments

This study was conducted with the support of a Dwight David Eisenhower Graduate Transportation Fellowship from the U.S. Department of Transportation and a Leslie E. and Stella M. Ellithorpe Student Research Award from Kansas State University. The authors thank Perla Reyes, Matthew Sanderson, and John Keller, all of Kansas State University, for providing guidance and reviewing drafts.

Author Contributions

The authors confirm contribution to the paper as follows: study conception and design: ER, GN; data collection: ER; analysis and interpretation of results: ER; draft manuscript preparation: ER, GN. All authors reviewed the results and approved the final version of the manuscript.

References

1. De Jong, G., J. Fox, A. Daly, M. Pieters, and R. Smit. Comparison of Car Ownership Models. *Transport Reviews*, Vol. 24, No. 4, 2004, pp. 379–408. https://doi.org/10.1080/0144164032000138733.
2. Sallis, J. F., M. F. Floyd, D. A. Rodríguez, and B. E. Saelens. Role of Built Environments in Physical Activity, Obesity, and Cardiovascular Disease. *Circulation*, Vol. 125, No. 5, 2012, pp. 729–737. https://doi.org/10.1161/CIRCULATIONAHA.110.969022.
3. Wachs, M. *Linkages Between Transportation Planning and the Environment*. Federal Reserve Bank of St. Louis, St. Louis, United States, 2000.
4. Nickerson, C., R. Ebel, A. Borchers, and F. Carriazo. *Major Uses of Land in the United States, 2007*. Publication EIB-89. U.S. Department of Agriculture, Economic Research Service, 2011.
5. US Census Bureau. 2010 Census Urban and Rural Classification and Urban Area Criteria. http://www.census.gov/geo/reference/ua/urban-rural-2010.html. Accessed June 23, 2016.
6. Holtzclaw, J., R. Clear, H. Dittmar, D. Goldstein, and P. Haas. Location Efficiency: Neighborhood and Socio-Economic Characteristics Determine Auto Ownership and Use - Studies in Chicago, Los Angeles and San Francisco. *Transportation Planning and Technology*, Vol. 25, No. 1, 2002, pp. 1–27. https://doi.org/10.1080/03081060290032033.
7. Caulfield, B. An Examination of the Factors That Impact upon Multiple Vehicle Ownership: The Case of Dublin, Ireland. *Transport Policy*, Vol. 19, No. 1, 2012, pp. 132–138. https://doi.org/10.1016/j.tranpol.2011.08.003.
8. Lee, S. S., and M. L. Senior. Do Light Rail Services Discourage Car Ownership and Use? Evidence from Census Data for Four English Cities. *Journal of Transport Geography*, Vol. 29, 2013, pp. 11–23. https://doi.org/10.1016/j.jtrangeo.2012.12.002.
9. Haas, P. M., G. L. Newmark, and T. R. Morrison. Untangling Housing Cost and Transportion Interactions: The Location Affordability Index Model—Version 2 (LAIM2). *Housing Policy Debate*, 2016, pp. 1–15. https://doi.org/10.1080/10511482.2016.1158199.
10. Sun, Y., E. Waygood, and Z. Huang. Do Automobility Cohorts Exist in Urban Travel?: Case Study of Osaka Metropolitan Area, Japan. *Transportation Research Record: Journal of the Transportation Research Board*, 2012. 2323: 18–24.
11. Pyddoke, R., and C. Creutzer. *Household Car Ownership in Urban and Rural Areas in Sweden 1999–2008*. Publication 2014:21. CTS - Centre for Transport Studies Stockholm (KTH and VTI), 2014.

12. Dargay, J. M. Determinants of Car Ownership in Rural and Urban Areas: A Pseudo-Panel Analysis. *Transportation Research Part E: Logistics and Transportation Review*, Vol. 38, No. 5, 2002, pp. 351–366. https://doi.org/10.1016/S1366-5545(01)00019-9.
13. Jovic, J., and B. Rankovic Plazinic. Transportation Demand Management in Rural Areas. Presented at the Towards a Humane City Conference, Novi Sad, Serbia, 2013.
14. Goodwin, R., J. Overman, and D. Rosa. *Rural Transportation Planning Guidebook*. Publication FHWA/TX-04/0-4230-P1. Center for Transportation Training and Research, Texas Southern University, Houston, Texas, 2004.
15. Dargay, J., and M. Hanly. Volatility of Car Ownership, Commuting Mode and Time in the UK. *Transportation Research Part A: Policy and Practice*, Vol. 41, No. 10, 2007, pp. 934–948. https://doi.org/10.1016/j.tra.2007.05.003.
16. Flamm, B. The Impacts of Environmental Knowledge and Attitudes on Vehicle Ownership and Use. *Transportation Research Part D: Transport and Environment*, Vol. 14, No. 4, 2009, pp. 272–279. https://doi.org/10.1016/j.trd.2009.02.003.
17. Schimek, P. Household Motor Vehicle Ownership and Use: How Much Does Residential Density Matter? *Transportation Research Record: Journal of the Transportation Research Board*, 1996. 1552: 120–125.
18. Oakil, A. T. M., D. Ettema, T. Arentze, and H. Timmermans. Changing Household Car Ownership Level and Life Cycle Events: An Action in Anticipation or an Action on Occurrence. *Transportation*, Vol. 41, No. 4, 2013, pp. 889–904. https://doi.org/10.1007/s11116-013-9507-0.
19. Giuliano, G., and J. Dargay. Car Ownership, Travel and Land Use: A Comparison of the US and Great Britain. *Transportation Research Part A: Policy and Practice*, Vol. 40, No. 2, 2006, pp. 106–124. https://doi.org/10.1016/j.tra.2005.03.002.
20. Cao, J., and X. Cao. The Impacts of LRT, Neighbourhood Characteristics, and Self-Selection on Auto Ownership: Evidence from Minneapolis-St. Paul. *Urban Studies*, Vol. 51, No. 10, 2014, pp. 2068–2087. https://doi.org/10.1177/0042098013505887.
21. Thakuriah, P., S. Menchu, and L. Tang. Car Ownership Among Young Adults: Generational and Period-Specific Perspective. *Transportation Research Record: Journal of the Transportation Research Board*, 2010. 2156: 1–8.
22. Clark, S. D. Mapping Car Ownership in Great Britain over Four Decades. *Journal of Maps*, Vol. 11, No. 2, 2015, pp. 354–360. https://doi.org/10.1080/17445647.2014.960484.
23. Cervero, R., and R. Gorham. Commuting in Transit versus Automobile Neighborhoods. *American Planning Association. Journal of the American Planning Association; Chicago*, Vol. 61, No. 2, 1995, p. 210. https://doi.org/10.1080/01944369508975634.
24. Khattak, A. J., and D. Rodriguez. Travel Behavior in Neo-Traditional Neighborhood Developments: A Case Study in USA. *Transportation Research Part A: Policy and Practice*, Vol. 39, No. 6, 2005, pp. 481–500. https://doi.org/10.1016/j.tra.2005.02.009.
25. Federal Highway Administration. Planning for Transportation in Rural Areas. https://www.fhwa.dot.gov/planning/publications/rural_areas_planning/ruralguide.pdf. Accessed February 6, 2017.
26. Federal Highway Administration. Small Town and Rural Multimodal Networks. https://www.fhwa.dot.gov/environment/bicycle_pedestrian/publications/small_towns/fhwa-hep17024_lg.pdf. Accessed February 6, 2017.
27. US Department of Agriculture Rural Development. Farm Bill | USDA Rural Development. http://www.rd.usda.gov/about-rd/farm-bill. Accessed June 12, 2016.
28. Minnesota Population Center. 11.0. *National Historical Geographic Information System*. University of Minnesota, Minneapolis, 2016.
29. US Census Bureau. American Community Survey (ACS). https://www.census.gov/programs-surveys/acs. Accessed April 7, 2017.
30. City of Decatur, Indiana Comprehensive Plan of 2010. http://www.decaturin.org/Documents/Comprehensive%20Plan_lowres.pdf. Accessed February 23, 2017.
31. Grammich, Clifford, Hadaway, Kirk, Houseal, Richard, Jones, Dale E., Krindatch, Alexei, Stanley, Richie, and Taylor, Richard H. Amish Groups, Undifferentiated Counties. *The Association of Religion Data Archives*, 2010. http://www.thearda.com/ql2010/QL_C_2010_2_979c.asp. Accessed April 6, 2017.
32. Yates Transit Service, Inc. Yates Transit Service. *Yates Transit Service*. http://yatestransitservice.com/home.html. Accessed April 6, 2017.
33. Greene County, New York. Greene County Transit Announces Expanded Routes & Schedule. *Greene County, New York*. http://greenegovernment.com/news-press/greene-county-transit-announces-enhanced-routes-schedule. Accessed April 6, 2017.
34. Trailways. Bus Service Locations in New York. *Trailways*. https://webstore.trailways.com/locations.asp?state=NY. Accessed March 30, 2017.
35. Grammich, Clifford, Hadaway, Kirk, Houseal, Richard, Jones, Dale E., Krindatch, Alexei, Stanley, Richie, and Taylor, Richard H. Yates County, New York. *The Association of Religion Data Archives*. http://www.thearda.com/ql2010/QL_C_2010_2_979c.asp. Accessed April 6, 2017.
36. Chouteau County. Chouteau County Planning/Development Regulations. *Chouteau County, Montana*. http://www.co.chouteau.mt.us/pages/development.htm. Accessed March 30, 2017.
37. Grammich, Clifford, Hadaway, Kirk, Houseal, Richard, Jones, Dale E., Krindatch, Alexei, Stanley, Richie, and Taylor, Richard H. Hutterian Brethren Counties. *The Association of Religion Data Archives*. http://www.thearda.com/ql2010/QL_C_2010_2_979c.asp. Accessed April 6, 2017.
38. Hutterites.org. Frequently Asked Questions about Hutterites. *Hutterian Brethren*. http://www.hutterites.org/day-to-day/faqs/. Accessed March 30, 2017.

The Standing Committee on Social and Economic Factors of Transportation (ADD20) peer-reviewed this paper (18-01545).

Article

TRR
JOURNAL OF THE TRANSPORTATION RESEARCH BOARD

Transportation Research Record
2018, Vol. 2672(6) 11–20

DOI: 10.1177/0361198118796738
journals.sagepub.com/home/trr
SAGE

Does Investing in Rail Transit Benefit the Poor? A Comparative Study of Rail and Bus Travel by Low-Income Households in the California Household Travel Survey

Ryland Lu[1]

Abstract

This paper addresses academic discourse that critiques urban rail transit projects for their regressive impacts on the poor and proposes bus funding as a more equitable investment for urban transit agencies. The author analyzed data from the 2012 California Household Travel Survey on transit trips in Los Angeles County. The author cross-tabulated data on the modal breakdown of transit trips by household income category and on the breakdown of household income associated with trips by bus and rail transit modes. The author also comparatively evaluated the speed of trips (as a ratio of miles per hour) taken by rail and by bus by low-income households in the county. The author found convincing evidence that, on average, trips low-income households made by rail transit covered a greater distance per hour than trips taken by bus transit, but that trips made on the county's bus rapid transit services with dedicated rights-of-way had a higher mean speed than those taken by rail. Moreover, the mode and income cross-tabulations indicate that rail transit projects only partially serve low-income households' travel needs. To the extent that equitable transit planning entails minimizing the disparities in access, both rail and bus rapid transit projects can advance social justice if they are targeted at corridors where they can serve travel demand by low-income, transit dependent households.

Since 1970, more than 20 US cities have built urban rail transit systems in an attempt to reduce the negative externalities of automobile dominance (*1*). In Los Angeles, efforts to build a regional heavy and light rail network in the 1990s encountered opposition from advocacy groups like the Bus Riders Union, who argued that a costly rail network would hurt low-income populations by diverting funds from bus operations (*2*). Historically, academic scholarship has tended to adopt this critical discourse and has broadly criticized postwar rail projects for benefitting affluent suburban "choice" riders at the expense of low-income, transit-dependent bus riders. However, survey data from the past few decades show that urban heavy and light rail ridership is not necessarily more affluent than bus ridership, while recent studies by scholars like Fan, Levinson and Guthrie indicate that the speed of new rail systems potentially improves low-income households' access to retail and jobs (*3*). Using data from the California Household Travel Survey, this paper addresses this gap in perspectives by assessing the extent to which low-income households in Los Angeles County utilize rail transit (for transit trips) and evaluating whether rail transit trips cover a greater distance per unit of time than bus trips.

Literature Review

Equity-focused scholarship has historically criticized rail transit projects on the grounds that they benefit the affluent at the expense of the poor. Pucher, Taylor and Garrett, Wachs and Taylor, Richmond, Blumenberg and Ong and Giuliano all characterize new rail projects as serving the interests of white and middle-income "choice" riders, who have access to automobiles, rather than the poor and minority riders who depend on transit because of their low incomes (*4–9*). More recently, Taylor and Morris argued for a dichotomy between poor and non-white bus riders and middle-class white train ridership, on the basis of three decades' worth of data from the National Household Transportation Survey (*10*). Taylor and Morris, Richmond, and Blumenberg and Ong argue that operational investments in bus services, such as dedicated bus lanes, priority signaling and increased bus

[1]Department of Urban and Regional Planning, UCLA Luskin School of Public Affairs, Beverly Hills, CA

Corresponding Author:
Address correspondence to Ryland Lu: rylandlu@g.ucla.edu

headways are more equitable for the poor than rail transit construction (*7*, *8*, *10*).

One flaw with this discourse is that it ignores the nuanced differentiation between rail and bus rider demographics that appears in national travel studies. Early in the era of postwar rail construction, Pucher's analysis of the 1977 Nationwide Personal Transportation Survey showed that although bus and streetcar riders (which comprised a single category in the survey) had lower incomes than subway riders, both demographics were significantly poorer than either commuter rail riders or drivers (*11*). Almost half of bus and streetcar riders and a third of subway riders came from the lowest two income brackets, compared to 17% of drivers and 15% of commuter rail riders (*11*). More recently, Pucher and Renne's analysis of the 2001 National Household Travel Survey (NHTS) revealed that, as a commuting mode, metro/subway/heavy Rail had a "bipolar distribution" of use among income groups, with the greatest usage by the lowest and highest income brackets (*12*): by contrast, the lowest income bracket had the highest share of Light Rail/Bus ridership. The geographic location of Los Angeles's heavy rail lines and several light rail lines in lower-income urban and suburban areas could likely result in a less affluent rail rider population than suggested by national averages (*2*).

Moreover, recent literature indicates that the speed of rail rapid transit systems may crucially benefit low-income workers, by improving their access to jobs. Sanchez, who evaluates the impact of transit accessibility on labor market participation in Portland and Atlanta, finds a correlation in the latter city between distance to either a bus or rail stop and hours worked, although he notes that distance to a rail stop has a lower correlation to the second variable than distance to a bus stop (*13*). Later studies have provided more affirmative evidence. Conducting geographic information system (GIS) analyses of nine Southern and Western cities (including Los Angeles), Griffin and Sener noted that cities with integrated bus and rail transit systems have lower average discrepancies between the percentages of low-income and total workers accessible by transit to a census block, compared to cities with only bus systems (*14*). Conversely, Fan, Guthrie and Levinson found that workers experienced increased access to jobs of all income levels following the opening of Minneapolis's Hiawatha light rail line, both in census blocks directly adjoining the light rail line and along connecting bus routes (*3*). Dividing the light rail line's course into four geographic areas, the authors observed that the level of accessibility increased the most for low-income jobs in two of the four geographic areas.

As evidenced in the papers discussed above, "accessibility" is most commonly defined in a transportation planning context as the number of opportunities available to an individual within a given distance or travel time (*15*). This "cumulative opportunity" measure of accessibility factors in both the travel time to opportunity sites from an individual's location (or the individual's "impedance") and the distribution of opportunity sites around his or her location (*16*). Although rail transit can potentially improve accessibility in an area through stimulating development densification—which increases the number of opportunity sites in a given space—the author finds the impedance component more pertinent for the present study because it directly relates to the mode's technology and is frequently perceived as a core benefit by rail transit's proponents (*7*). While not a direct measure of accessibility, analyzing the per-mile travel time for rail trips taken by low-income households in Los Angeles County (and comparing this to the per-mile travel time for bus and bus rapid transit trips taken by these households) may indicate an important benefit of rail transit for low-income travelers.

Survey Analysis

Methodology

This study involved a comparative analysis of the set of bus and rail transit trips in Caltrans' 2012 California Household Travel Survey that were taken by low-income households in Los Angeles County (*17*). The California Household Travel Survey collected information on daily trips made by 42,431 households across the state between January 2012 and March 2013 (*17*). The agency recruited households for the survey using a stratified random sampling technique that accounted for socioeconomic- and geographic-based underrepresentation (*17*). In addition to gathering information on each trip made by a person in a recruited household during the period (*17*), the survey obtained demographic and socioeconomic data on all participating households (including county of residence, household size, and household income), as well as information on each household's vehicles.

Roughly 36,000 households participating in the survey (86.5% of total participants) recorded all daily trip data (including the variables outlined in Table 1) in travel diaries and either mailed the diaries to Caltrans or input data online or through a computer-assisted phone interview (*17*). Trip information entered online or over the phone was processed using TripBuilder software, that geocoded trip origins and destinations and then assessed the most logical travel route using data from Google Transit, for transit trips. 5,199 households (13.4% of the total) had their vehicle trips tracked using a global positioning system (GPS) device, and recorded information on other trips using the diaries (*17*).

One objective of the analysis was to evaluate whether bus or rail transit better serves the travel needs of Los

Table 1. Variables of Analysis from the California Household Travel Survey

Data set	Variable description	Variable name	Categories examined for study
Households	Sample Number	SAMPN	
Households	Residential County	CFTIP	06037 = LOS ANGELES
Households	Household income	INCOM	1 = $ 0–9,999 2 = $10,000–24,999 3 = $25,000–34,999 4 = $35,000–49,999 5 = $50,000–74,999 6 = $75,000–99,999 7 = $100,000–149,999 8 = $150,000–199,999 9 = $200,000–249,999 10 = $250,000 or more 98 = DK 99=RF
Place (Trips)	Person number	PERNO	
Place (Trips)	Mode of trip	MODE	15 = Local bus, rapid bus 16 = express bus/commuter bus (AC Transbay, Golden Gate Transit, etc) *17 = Premium bus (Metro orange silver line)* 18 = School bus 19 = Public transit shuttle (DASH, Emery Go Round, etc.) 20 = AirBART/LAX FlyAway 21 = Dial-a-Ride/Paratransit (access services, etc.) 22 = Amtrak bus 23 = Other bus 24 = BART, Metro red/purple line 25 = ACE, Amtrak, Caltrain, Coaster, Metrolink 26 = Metro blue/green/gold line, Muni Metro, Sacramento light rail, San Diego Sprinter/trolley/orange/blue/green, VTA light rail 27 = Street car/cable car
Place (Trips)	Travel time to place in minutes	TRPDUR	
Place (Trips)	RouteDistance	TripDistance	
Place (Trips)	AirTripDistance	TRPDIST	
Place (Trips)	Trip location county	CTFIP	06037=LOS ANGELES

Note: Italicized text indicates only included in the final part of the analysis.
Source: 2010–2012 California Household Travel Survey.

Angeles County's low-income households. To this end, the author first cross-tabulated transit trip modes by income status, which delineated low-income and non-low-income categories. Frequency distributions of household income tier for households associated with bus trips and households associated with rail trips in the county were then obtained. These distributions grouped the survey's nine household income categories into three income tiers. The two analyses provided information on the transit modes low-income households most frequently rely on and on the economic profiles of bus and rail transit ridership.

The second component of the analysis involved a comparison of travel times for bus and rail transit trips taken by low-income households. The analysis specifically compared the statistical distribution of the variable of the trip speed (the ratio of the trip distance, measured in miles, to the trip duration, measured in minutes, multiplied by 60) for the samples of bus trips, rail trips and trips made by "premium bus" taken by low-income households. The analysis focused exclusively on trips made within Los Angeles County by households residing in the county.

The specific variables incorporated into the study are summarized (in the format presented in the survey) in Table 1. The bus trip types and rail trip types included in the analysis are highlighted in bold under the trip mode category classification. The household income categories comprising the definition used in this paper of low-income households are also highlighted.

The analysis of trip characteristics limited bus trips to the category of "Local and Rapid" bus trips (code 15), because these correspond to the primary types of conventional urban bus service in Los Angeles County. The rail transit category groups together heavy rail trips (code 24) and light rail trips (code 26), as the two modes comprise the Los Angeles County Metropolitan Transit Authority's (or "Metro's") rail system that serves the central urban neighborhoods of Los Angeles County. In the analysis of trip speed, trips coded as "premium bus", a designation that applies to Metro's two bus rapid transit lines (orange and silver lines) that employ dedicated rights-of-way and pre-boarding fare payment (*18*), comprised a third category alongside bus and rail. Including the Premium Bus trip category permitted an examination of the effects of rail-like improvements to the bus service on trip speed.

The survey gives two variables of trip distance: "trip distance" uses the data obtained from the TripBuilder software (*17*), and "airtrip distance" gives the linear (or air) distance between a trip origin and destination. The former measure ostensibly gives the actual route traveled, but it seemed to err in a few cases. Based on the recommendation of a Caltrans representative, the author used a logic function in Microsoft Excel to solve the greater of the two variables and created a new variable from the results titled "DistanceFin." This last variable was used to calculate the ratio of trip speed.

The listing of income by category rather than numerical value in the survey, made it difficult to distinguish low-income from non-low-income households according to the federal government's poverty threshold, as initially desired, as these vary based on the number of persons in a household (*19*). This study used the survey's cut-off between income categories 3 and 4 ($35,000 a year) to define low-income households because it is the highest cut-off figure, which is less than 80% of Los Angeles County's 2014 median household income of around $55,870 per year (*20*). The latter figure is the cutoff for the Department of Housing and Urban Development's definition of a "lower-income household" for a family of four (*21*).

Caltrans delivered the survey data in a Microsoft Access database file. After running a query that linked information from the household demographic and trip databases, the author converted the database file to a Microsoft Excel spreadsheet document to conduct analysis in Stata.

Results

Transit Mode Breakdown by Income Category. The travel survey has a total of 1965 transit trips associated with low-income households and 1007 transit trips associated with non-low-income households within Los Angeles County. There were 422 households in the low-income group and 289 households in the non-low-income group. As Table 2 illustrates, the majority of trips taken by both low-income and non-low-income households were coded as conventional bus trips, although they comprised a smaller share of the total trips taken by non-low-income households (58% of trips as compared to 82% of trips for low-income households). Conversely, light rail and heavy rail trips comprised only about 12.3% of transit trips associated with low-income households, but almost 30% of transit trips associated with non-low-income households. Each of the other modes individually comprised of less than 4% of the total trips associated with households in both income categories, although the percentages of express bus trips and commuter rail trips associated with non-low-income households were almost double the percentages of express bus trips and commuter rail trips associated with low-income households.

However, the breakdowns of bus and rail transit trips by household income tier (Table 3), show that households in the low-income tier were associated with the largest share of both bus and rail transit trips, although they only accounted for a majority of trips by the former mode. For both mode categories, middle-income households (defined as households with incomes of $35,000–99,000 a year, corresponding to codes 4 through 6 on the survey) accounted for the second largest share of trips and upper-income households (defined as households with incomes over $99,000 a year, corresponding to

Table 2. Breakdown of Transit Trips in Los Angeles County by Mode and Income

Trip mode	Non-low-income households		Low-income households	
Local or rapid bus	589	58.40%	1590	81.91%
Express bus	37	3.60%	17	0.86%
Premium bus	26	2.58%	50	2.54%
DASH or public transit shuttle	19	2.38%	40	2.04%
Heavy rail (subway)	102	10.12%	64	3.26%
Commuter rail	32	3.18%	25	1.27%
Light rail	196	19.46%	177	9.00%
Total	1007	100.00%	1963	100.00%

Table 3. Breakdown of Household Income by Transit Trip Mode

MODE	Under $35,000	$35,000–99,000	Over $99,000	DK/RF	Total
Bus	1,610	458	145	205	2,418
	66.6%	18.9%	6.0%	8.5%	100.0%
Rail	241	205	96	31	573
	42.1%	35.8%	16.8%	5.4%	100.0%

Note: DK/RF = don't know/refuse to state.

Table 4. Statistical Distribution of Trip Duration by Transit Mode

MODE	Count	Min.	Mean	Median	Max.	SD
BRT	50	4	22.31592	24	58	12.76017
Bus	1590	1	21.15093	15	409	18.24848
Rail	241	1	18.8913	15	80	14.57103

Note: Min = minimum; max = maximum; SD = standard deviation; BRT = bus rapid transit.

Table 5. Statistical Distribution of Trip Distance by Transit Mode

MODE	Count	Min.	Mean	Median	Max.	SD
BRT	50	0.0498337	9.35959	10.36473	22.3332	5.272763
Bus	1590	0	4.248006	2.615651	108.829	5.096534
Rail	241	0	6.88181	5.175679	28.64496	5.772548

Note: Min = minimum; max = maximum; SD = standard deviation; BRT = bus rapid transit.

codes 7 through 10 on the survey) accounted for the smallest share of trips. However, middle-income and upper-income households accounted for almost twice as many rail trips as bus trips in the survey. A greater percentage of bus trips were associated with respondents who either didn't know or refused to state their income (codes 98 and 99 in the survey).

Trip Speed Analysis. Comparison of trip durations for bus and rail transit trips revealed little difference in the statistical distributions (see Table 4). The set of bus trips and the set of rail trips both had a median duration of 15 minutes, although the former set had a slightly higher mean duration figure than the latter. However, the set of bus rapid transit trips had a median duration of 24 minutes, with the mean figure slightly lower than the median.

In contrast, for trip distance (see Table 5), the median distance for rail trips was about double the median distance for bus trips, and the median trip distance for the set of bus rapid transit trips was about four times the median bus trip distance.

The distribution of the miles per hour statistic (Table 6) similarly shows that low-income respondents' trips by rail and bus rapid transit (premium bus) tended to cover a significantly greater distance per fixed duration than their trips by bus. The mean and median speeds for bus rapid transit trips were both greater than the mean and median speeds for rail trips.

A two-sample t-test shows that the t-statistic for the difference between the mean bus and rail trip speeds (Figure 1) is negative and significant at almost a 100% confidence level. This establishes with near certainty that the mean rail trip speed is greater than the mean bus trip speed among the population of low-income transit riders.

However, a t-test for the difference between the mean bus rapid transit and rail trip speeds (Figure 2) shows a positive t-statistic, that is significant at a 95% confidence level but not at a 99% confidence level. This strongly suggests that the population mean speed for trips taken by low-income households on the orange and silver bus rapid transit lines is greater than the population mean speed for subway and light rail trips taken by low-income households in Los Angeles County.

Shortcomings. It should be acknowledged that the survey does not provide a perfect measure of the travel activity of low-income transit riders. Despite its carefully-tailored random sampling, the study could be biased by the small

Table 6. Statistical Distribution of Trip Speed (Miles per Hour) by Mode

MODE	Count	Min.	Mean	Median	Max.	SD
BRT	50	0.7475053	28.31816	25.98389	70	14.97204
Bus	1590	0	13.2288	10.17368	70	11.95085
Rail	241	0	23.40815	21.49521	70	14.78677

Note: Min = minimum; max = maximum; SD = standard deviation; BRT = bus rapid transit.

```
Two-sample t test with unequal variances

   Group |     Obs        Mean    Std. Err.   Std. Dev.   [95% Conf. Interval]
---------+--------------------------------------------------------------------
     bus |    1590     13.2288    .2997094    11.95085    12.64093    13.81667
    rail |     241    23.40815    .9524993    14.78677    21.53182    25.28447
---------+--------------------------------------------------------------------
combined |    1831    14.56863    .2997715    12.82728     13.9807    15.15656
---------+--------------------------------------------------------------------
    diff |           -10.17935    .9985392               -12.14466   -8.214028
------------------------------------------------------------------------------
    diff = mean(bus) - mean(rail)                                 t = -10.1942
Ho: diff = 0                     Satterthwaite's degrees of freedom =  289.448

    Ha: diff < 0                 Ha: diff != 0                 Ha: diff > 0
 Pr(T < t) = 0.0000         Pr(|T| > |t|) = 0.0000          Pr(T > t) = 1.0000
```

Figure 1. Two sample *t*-test for the difference between the mean bus and rail speeds.

```
   Group |     Obs        Mean    Std. Err.   Std. Dev.   [95% Conf. Interval]
---------+--------------------------------------------------------------------
     brt |      50    28.31816    2.117366    14.97204    24.06315    32.57316
    rail |     241    23.40815    .9524993    14.78677    21.53182    25.28447
---------+--------------------------------------------------------------------
combined |     291    24.25179    .8739618    14.90867    22.53168     25.9719
---------+--------------------------------------------------------------------
    diff |            4.910011    2.302781                .3776633    9.442359
------------------------------------------------------------------------------
    diff = mean(brt) - mean(rail)                                 t =   2.1322
Ho: diff = 0                                     degrees of freedom =      289

    Ha: diff < 0                 Ha: diff != 0                 Ha: diff > 0
 Pr(T < t) = 0.9831         Pr(|T| > |t|) = 0.0338          Pr(T > t) = 0.0169
```

Figure 2. The *t*-test for the difference between the mean bus rapid transit and rail speeds.

percentage of sampled households that completed the survey. Statewide, only 4.9% of households selected by the sample were recruited for the survey and only 67% of recruited households completed the survey (*17*). A report on the survey notes a disproportionately low recruitment rate (4.8% versus 18% of the state population) from households that had only mobile phone service: this demographic comprises a rapidly increasing share of young, lower-income and mobile households (*17*). The set of transit trips in Los Angeles County profiled in the survey corresponded to only 711 households, raising further questions of representation. However, this figure's share of the total households sampled in Los Angeles County (7.1%) is equivalent to the percentage of workers in Los Angeles who commuted by transit according to the 2012 American Community Survey (Social Explorer2). The sets of low-income households associated with bus (376 households) and rail (79 households) transit trips—as well as the sets of bus and rail trips—still provide large enough samples (each having more than 50 participants [*23*]) to obtain a normal sampling distribution that can be used to approximate population statistics and accurately test for statistical significance. On the other hand, the 50 premium bus trips made by low-income households in Los Angeles County corresponded to only 13 households, making for a potentially unrepresentative sample of trips in this category.

In the survey there were 32 conventional bus trips, 3 premium bus trips and 5 rail trips cataloged that had unrealistically low trip duration figures for the distance recorded, which probably resulted from errors in respondents' diary recordings. As suggested by a Caltrans representative, the trip duration for these trips was revised (based on the distance traveled) so that the trip speed figure amounted to 70 miles per hour, a figure in between the typical top operating speed of around 55 miles per hour for bus (*24*) and light rail vehicles (*25*), and 80 miles per hour for heavy rail vehicles (*26*).

The data obtained from the survey by the author did not provide information on geographical units smaller than the census tract level. Access to data on census block location, as well as data on respondents' address and location of their origins and destinations, could only be obtained through signing a separate Confidentiality Agreement that Caltrans only approves for institutions engaged in a formal planning study. The author's lack of access to such data prevented independent verification of the trip distance and duration figures using the transit agencies' general transit feed specifications (GTFS). Although the author considered analysis of trip data at the census tract level (which could have provided some information on the distribution of bus and rail trip origins), a lack of proficiency in GIS at the time of the analysis was conducted and concern for the adequacy of sample sizes at this geographic level discouraged the author from conducting such an analysis.

Finally, individual trips in the survey are "unlinked" and do not account for trip chaining or use of multiple modes (M. Assadi, personal communication, November 2016). The analysis thus gives incomplete measures of trip speed, which disregard the distance that households might have to travel to access bus or rail transit.

Interpretation of Data

The first part of the study indicated that low-income transit riders in Los Angeles County take more trips by bus than by rail, while still making extensive use of the latter mode. Trips taken by low-income households in the travel survey comprised a large majority of total bus trips and a clear plurality of rail trips. By contrast, trips taken by highest household income category comprised only 15% of the total rail trips in Los Angeles County, indicating that rail is hardly the elite mode which its critics portray it as.

The finding that low-income households in the survey took more bus than rail trips, despite taking a large share of total rail trips, probably pertains to the greater coverage of bus transit in the county (*27*). Past National Household Travel Survey data shows that low-income households are less likely to own an automobile than households in higher income categories (*12*). Low-income households probably rely on the conventional bus service not only to access neighborhood destinations for which rail transit is geographically inconvenient, but also to travel between rail stations and trip origins and destinations: Many rail trips in the survey probably have at least two associated bus trips. On average, rail trips associated with low-income households (in the survey) have a similar duration to bus trips but extend a considerably greater distance, further suggesting that low-income households rely more heavily on bus transit for local and connecting trips and on rail transit for long-distance commuting.

The analysis of trip speed offered strong evidence that trips taken by rail transit are faster than trips taken by conventional bus transit for low-income transit riders in Los Angeles County, but slightly slower than trips made on the premium bus lines. This implies that the speed differential between rail transit and bus transit does not derive from an inherent technological advantage, and that bus services can compete with rail services for speed when provided with a dedicated right-of-way, station facilities (which speed service by providing facilities for prepaid ticketing [*24*]), and limited stops. Unfortunately, as mentioned previously, the small number of households associated with the set of premium bus trips make the statistics for this category susceptible to reflecting individual travel idiosyncrasies. Additionally, the premium bus category excludes less advanced bus rapid transit services like Metro's Rapid 720 bus along Wilshire Blvd (*28*), which provides limited stop service and receives signal priority but lacks a dedicated lane outside of peak hours (*29*). As Hess, Taylor and Yoh note, bus rapid transit services with dedicated lanes are the most expensive type of bus transit service to provide (*30*), with per-mile costs as high as $23 million on the orange line (*31*) – within $11 million of the national average cost for light rail projects in 2000. However, the orange line's costs still fall far below those of local light rail projects like the first phase of Metro's gold line, which costs roughly $60 million per mile (*32*). Although the results only apply to a limited sample of rides on the most advanced form of bus rapid transit technology, they suggest that bus rapid transit services with dedicated rights-of-way can offer enhanced mobility benefits for the poor at a lower cost than rail.

To the extent that both rail and bus rapid transit reduce travel times for low-income transit riders, they could have a highly positive impact, by increasing the predictability of riders' commutes to work and other activities (*33*). Based on Handy and Niemeier's definition, however, a conclusive assessment of whether rail or bus rapid transit enhances accessibility in Los Angeles will also require an examination of the number and type of destinations that the modes service (*16*). The California Household Travel Survey's lack of available data on geographic origins and destinations limits its usefulness for such a study.

Discussion and Conclusion

The study's findings indicate that both rail and bus rapid transit offer substantial benefits to low-income households in Los Angeles County through reducing travel time costs. Given the limitations of the survey data and the variables that were analyzed, both findings should be interpreted cautiously. To the extent these findings hold true, they suggest, on the one hand, that planners at the Los Angeles County Metro should more frequently consider implementing bus rapid transit in place of rail on corridors where the agency can obtain a dedicated right-of-way. In corridors where the agency could fund more expansive capital projects with the former mode due to a lower per-mile cost (as has been the case in Latin American cities like Quito [*24*]), it would offer the greatest mobility dividend for low-income households that rely on transit.

At the same time, transit agencies like Metro should place more emphasis on serving low-income households and travel destinations with rapid transit investments. Elkind documents how, in the particular case of Los Angeles County, political appeasement of affluent suburban interests (who are represented on the boards of Metro and predecessor regional public transportation agencies) has shaped the development of rail transit system, with extensions of the gold line to Pasadena and Azusa, for instance, prioritized over the development of the Wilshire Subway (*34*). This allocation (which Nelson and Sanchez suggest as being common for agencies with suburban representation [*35*]) is doubly tragic because lower-income urban neighborhoods in Los Angeles are more likely than average to have the density needed to justify costly transit investments (*9*). Martens, Golub and Robertson argue (based on Rawls' Theory of Justice) that socially-just transit planning should strive to maximize the average access level for a geographic unit, while minimizing the gaps in access between car-owning and car-less populations within neighborhoods (*36*). Investing in rapid transit projects that reduce the impedance in trips taken by transit-dependent, low-income individuals relative to that in automobile trips (lessening the need to trade time costs for the monetary costs of automobile ownership) can fit such a framework of planning justice.

Accordingly, Metro's long-range transit planning should prioritize projects such as the proposed Vermont Avenue busway, that serve corridors with high volumes of bus ridership and greater-than-average percentages of low-income or transit-dependent riders (*37*). Equity-focused rail and bus rapid transit advocates can potentially learn from the Bus Riders Union (*2*), and employ social justice rhetoric and broad-based coalition building to motivate grassroots and legal activism for planning transit investments that reduce access inequities.

In addition, the findings highlight that rail or bus rapid transit capital projects should not come at the expense of a conventional bus service. Low-income households in Los Angeles County continue to rely on conventional bus services for the bulk of trips even when they travel quickly and frequently by rail or bus rapid transit. Fan, Guthrie and Levinson suggest that increases in accessibility associated with light rail expansion also accrue to the connecting bus lines (*3*). Therefore, ensuring adequate frequency of connecting bus services at transit stations can help low-income riders capitalize off time savings from rail and bus rapid transit. Agencies like Metro should also invest in signal timing and boarding improvements to speed bus operations on corridors that do not merit the density for exclusive infrastructure.

Given the costs of car ownership, the quality of transit service is an important determinant of the mobility and economic well-being for low-income households (*9*). To the extent that well-designed rail transit—operating in tandem with local bus service and bus rapid transit—can potentially reduce travel times for low-income households in Los Angeles County, advocates for transportation justice should fight against inequity in transit funding and construction rather than on behalf of a particular transit mode.

Acknowledgments

I thank the California Department of Transportation for providing access to the 2012–2013 California Household Travel Survey results. I am especially grateful to Mohammad Assadi at the department who took the time to answer all the questions I had on interpreting the survey data.

Author Contributions

The author confirms sole responsibility for the following: study conception and design, data collection, analysis and interpretation of results, and manuscript preparation.

References

1. Guerra, E., and R. Cervero. Transit and the "D" Word. *ACCESS Magazine*, Vol. 1, No. 40, 2012.
2. Grengs, J. Community-Based Planning as a Source of Political Change: The Transit Equity Movement of Los Angeles' Bus Riders Union. *Journal of the American Planning Association*, Vol. 68, No. 2, 2002, pp. 165–178.

3. Fan, Y., A. Guthrie, and D. M. Levinson. Impact of Light Rail Implementation on Labor Market Accessibility: A Transportation Equity Perspective. *Journal of Transport and Land Use*, Vol. 5, No. 3, 2012, pp. 28–39.
4. Pucher, J. Discrimination in Mass Transit. *Journal of the American Planning Association*, Vol. 48, No. 3, 1982, pp. 315–326.
5. Taylor, B. D., and M. Garrett. Reconsidering Social Equity in Public Transit. *University of California Transportation Center*, 1999, pp. 6–27.
6. Wachs, M., and B. D. Taylor. Can Transportation Strategies Help Meet the Welfare Challenge? *Journal of the American Planning Association*, Vol. 64, No. 1, 1998, pp. 15–19.
7. Richmond, J. E. D. The Mythical Conception of Rail Transit in Los Angeles. *Journal of Architectural and Planning Research*, Vol. 15, 1998, pp. 294–320.
8. Blumenberg, E., and P. Ong. Cars, Buses, and Jobs: Welfare Participants and Employment Access in Los Angeles. *Transportation Research Record: Journal of the Transportation Research Board*, 2001. 1756: 22–31.
9. Giuliano, G. Low Income, Public Transit, and Mobility. *Transportation Research Record: Journal of the Transportation Research Board*, 2005. 1927: 63–70.
10. Taylor, B. D., and E. A. Morris. Public Transportation Objectives and Rider Demographics: Are Transit's Priorities Poor, Public Policy? *Transportation*, Vol. 42, No. 2, 2014, pp. 347–367.
11. Pucher, J. Equity in Transit Finance: Distribution of Transit Subsidy Benefits and Costs Among Income Classes. *Journal of the American Planning Association*, Vol. 47, No. 4, 1981, pp. 387–407.
12. Pucher, J., and J. L. Renne. Socioeconomics of Urban Travel: Evidence from the 2001 NHTS. *Transportation Quarterly*, Vol. 57, No. 3, 2003, pp. 49–77.
13. Sanchez, T. W. The Connection between Public Transit and Employment: The Cases of Portland and Atlanta. *Journal of the American Planning Association*, Vol. 65, No. 3, 1999, pp. 284–296.
14. Griffin, G. P., and I. N. Sener. Equity Analysis of Transit Service in Large Auto-Oriented Cities in the United States. Presented at 94th Annual Meeting of the Transportation Research Board, Washington, D.C., 2015.
15. Hanson, S. The Context of Urban Travel. *The Geography of Urban Transportation*, 3rd Edition, 2004, pp. 3–29.
16. Handy, S. L., and D. A. Niemeier. Measuring Accessibility: An Exploration of Issues and Alternatives. *Environment and Planning A*, Vol. 29, No. 7, 1994, pp. 1175–1194.
17. NUStats Research Solutions. *2010-2012 California Household Travel Survey Final Report Appendix*. California Department of Transportation, Sacramento, CA, 2013.
18. *2016 Metro Transit Service Policies and Standards*. Los Angeles County Metropolitan Transit Authority, Los Angeles, CA, 2015. https://media.metro.net/images/service_changes_transit_service_policy.pdf. Accessed November 15, 2017.
19. Data. *Poverty Thresholds*. United States Census Bureau, Washington, D.C. https://www.census.gov/data/tables/time-series/demo/income-poverty/historical-poverty-thresholds.html. Accessed August 1, 2017.
20. United States Census Bureau. *Quick Facts: Los Angeles County*. https://www.census.gov/quickfacts/fact/table/losangelescountycalifornia/INC110215. Accessed August 1, 2017.
21. U.S. Department of Housing and Urban Development Office of Policy Development & Research. *FY 2016 HUD Income Limits Briefing Material*. U.S. Department of Housing and Urban Development, Washington, D.C., 2016. https://www.huduser.gov/portal/datasets/il/il16/IncomeLimitsBriefingMaterial-FY16.pdf
22. United States Census Bureau/American Fact Finder. B08006: Sex of Workers by Means of Transportation to Work. 2008-2012 American Community Survey, 2012. https://factfinder.census.gov/faces/tableservices/jsf/pages/productview.xhtml?pid=ACS_12_5YR_B08006&prodType9=table
23. Frankfort-Nachmias, C., and A. Leon-Guerrero. *Social Statistics for a Diverse Society*, 7th ed. SAGE Publications, Thousand Oaks, CA, 2015.
24. Cervero, R. *Bus Rapid Transit (BRT): An Efficient and Competitive Mode of Public Transport*. Berkeley Institute of Urban and Regional Development, Berkeley, CA, Working Paper 2013-1, 2013. http://iurd.berkeley.edu/wp/2013-01.pdf
25. *Commuter Rail vs. Light Rail*. RTD Fastracks, Regional Transportation District of Denver (RTD), Denver, CO, 2007–2016. http://www.rtd-fastracks.com/main_398. Accessed August 1, 2017.
26. Johnson, M. Average Schedule Speed: How Does Metro Compare? *Greater Greater Washington*. March 16, 2010. https://ggwash.org/view/4524/average-schedule-speed-how-does-metro-compare . Accessed August 1, 2017.
27. *Bus and Rail System*. Metro.net, June 2016. http://media.metro.net/riding_metro/maps/images/system_map_2016-0520.pdf. Accessed August 1, 2017.
28. *Route Map: 720*. Metro.net, 2017. https://media.metro.net/riding_metro/bus_overview/images/720.pdf. Accessed November 15, 2017.
29. Lemon, J. Longest Segment of Wilshire Peak Hour Bus Lanes to Open on Wednesday. *The Source*. April 7, 2015. http://thesource.metro.net/2015/04/07/longest-segment-of-wilshire-brt-bus-lanes-set-to-open-on-wednesday/. Accessed November 15, 2017.
30. Hess, D., B. Taylor, and A. Yoh. Light Rail Lite or Cost-Effective Improvements to Bus Service? Evaluating Costs of Implementing Bus Rapid Transit. *Transportation Research Record: Journal of the Transportation Research Board*, 2005. 1927: 22–30.
31. Rudlick, R. The Myth of the Magic Bus: The Weird Politics and Persistently Strange Logic Behind the Orange Line. *Streetsblog LA*. May 27, 2014. http://la.streetsblog.org/2014/05/27/the-myth-of-the-magic-bus-the-weird-politics-and-persistently-strange-logic-behind-the-orange-line/. Accessed November 29, 2016.
32. Loukaitou-Sideris, A., D. Houston, and A. Bromberg. Gold Line Corridor Study: Final Report. Ralph and Goldy Lewis Center for Regional Policy Studies, March 2007. http://www.reconnectingamerica.org/assets/Uploads/2007GoldLineCorridorStudy.pdf. Accessed August 1, 2017.

33. Blumenberg, E., and M. Manville. Beyond the Spatial Mismatch: Welfare Recipients and Transportation Policy. *Journal of Planning Literature*, Vol. 19, No. 2, 2004, pp. 182–205.
34. Elkind, E. *Railtown: The Fight for the Los Angeles Metro Rail and the Future of the City*. University of California Press, Berkeley, CA, 2014. http://www.sciencedirect.com/science/article/pii/S0967070X14002145#bib47
35. Nelson, A., T. Sanchez, J. Wolf, and M. Farquhar. Metropolitan Planning Organization Voting Structure and Transit Investment Bias: Preliminary Analysis with Social Equity Implications. *Transportation Research Record: Journal of the Transportation Research Board*, 2004. 1895: 1–7.
36. Martens, K., A. Golub, and G. Robinson. A Justice-theoretic Approach to the Distribution of Transportation Benefits: Implications for Transportation Planning Practice in the United States. *Transportation Research Part A*, Vol. 46, No. 4, 2012, pp. 684–695.
37. Hymon, S. Measure M: Bus Rapid Transit on Vermont Avenue. *The Source*, May 4, 2016. http://thesource.metro.net/2016/05/04/potential-ballot-measure-bus-rapid-transit-on-vermont-avenue-in-l-a/comment-page-2/. Accessed August 1, 2017.

The Standing Committee on Social and Economic Factors of Transportation (ADD20) peer-reviewed this paper (18-06719).

Article

TRR
JOURNAL OF THE TRANSPORTATION RESEARCH BOARD

Transportation Research Record
2018, Vol. 2672(6) 21–30

DOI: 10.1177/0361198118774153
journals.sagepub.com/home/trr
SAGE

Spatio-Temporal Hedonic Price Model to Investigate the Dynamics of Housing Prices in Contexts of Urban Form and Transportation Services in Toronto

Jason Hawkins[1] and Khandker Nurul Habib[1]

Abstract

A spatio-temporal hedonic price model is developed for the Greater Toronto area to examine the effects of urban configurations and proximity to transit services on housing price. A spatial Durbin panel model is utilized to account for both spatial and temporal autocorrelation. This model is shown to have advantages through its ability to reduce the number of explanatory variables required to obtain a strong fit with empirical data. Analysis is completed for the period of 1996 to 2017 and distinctions are made in housing stock between single-family houses, townhouses, and condominiums. It is shown that heterogeneities exist between the hedonic representations of each dwelling type and that separate models should be employed for each. In all cases, the average income of the community, its distance to the central business district (CBD), and population and employment density are found to be significant factors in the determination of price.

The Greater Toronto Area (GTA) is one of the hottest real estate markets in Canada. The average sale price of a house hit a record high of $920,791 in April 2017 (*1*). This is happening in the context of a growing disparity between housing price and the real income of residents of the GTA. According to a 2013 study conducted by the Royal Bank of Canada and Pembina Institute, the real price of housing in the GTA increased by 80% over the 30 years prior to 2012, while real income only increased by 18% (*2*). Numerous studies have found that the transportation infrastructure and the built environment influence the price of housing (*3–6*). As such, the GTA is an ideal candidate for investigating how policies relating to urban configuration and transportation services can influence housing price. We believe the results of such an investigation can help in the development of innovative forms of residential price modelling given the combined influences of changing housing dynamics and major changes depicted in the regional transportation vision, The Big Move (*7*). However, before embarking on this investigation, we reviewed the literature that presented methods and results for similar investigations.

Two previous studies of housing price have been conducted for the GTA (*8*, *9*). In general, there are two features of these studies that are pertinent to the present investigation. It is a standard method in the hedonic price literature to normalize price, typically by taking its natural logarithm. This has been found to produce a better fit (*9*), suggesting a non-linear relationship between price and the components of price encapsulated by explanatory variables. Habib used Box–Cox transformations to generalize this method, which proved effective in improving model fit (*9*). Spatio-temporal variations were introduced through a multi-level model specification. At the first level, dwellings were nested within a spatial cluster representing the neighbourhood. The temporal dimension was introduced at the second level via temporal clustering. A good fit was produced from the model, but Habib (*9*) suggested further analysis should be conducted to explore how coefficients vary over both space and time through the introduction of a third level with spatio-temporal clusters.

In addition, the majority of residential price models are developed at the level of individual dwellings and include sales data over a period of years. This has advantages in that dwelling-specific characteristics can be examined and many explanatory variables can be included in the model specification. The first challenge

[1]Transportation Engineering and Planning, Department of Civil Engineering, University of Toronto, Toronto, ON, Canada

Corresponding Author:
Address correspondence to Jason Hawkins: jason.hawkins@mail.utoronto.ca

associated with applying a spatial model to individual property data arises in the specification of the weight matrix. This is used in the identification of adjacent properties, which it is assumed have some influence on the dependent variable, the – sale price. Diao and Ferreira assumed a 400 meter band around each property within which other properties affect its price, while Pace et al. assumed the nearest 15 property sales in any prior year influence current property prices (*4, 10*). This value was based on the maximization of model fit. Other challenges arise in these models during the explanatory variable choice process. Despite the availability of detailed price data, many variables including household income, remain as averages from zone-based census data. Another example of the short-comings of detailed hedonic models, with many explanatory variables, is illustrated by the models of Diao and Ferreira and Otto and Schmid (*4, 11*). Both consider the effect of school quality on the price of housing, a challenging and controversial variable to quantify. However, spatial econometric theory continues to develop methods for describing spatial and temporal effects on prices, without the introduction of questionable variables.

Otto and Schmid applied advanced spatial dynamic panel data (SDPD) models to average single-family house prices in Germany across 412 administrative districts (*11*). They found this structure performed better than a purely spatial model, suggesting temporal effects are significant. Dwelling price was found to be strongly correlated with population density and per capita real disposable income in urban areas (*11*). Incorporating transportation considerations, a study by Iacono and Levinson found employment accessibility is an important component captured in property price (*12*). They used the number of employment opportunities within a 30 minute drive of the dwelling as a measure of accessibility. This was found to be relatively inelastic, with a 10% increase in the measure raising the sale price by an average of 2.3% (*12*). They also considered a measure of population access to ascertain the effect of competition for employment on sale price. This had a negative association with property price, with a 10% increase in the number of competing workers inducing an average 1% decline in price.

None of the models found in the literature included a distinction between dwelling structure type. Analysis was principally focused on single-family houses, with minimal mention of other dwelling types. Single-family house sales represent the largest source of residential sales, but other dwelling types are not insignificant and warrant analysis. In the case of our study area, increasing costs for home ownership may lead to changes in dwelling type choice and a transition to condominiums and townhouses. By including models for each of these dwelling types, we can compare the similarities and differences in the hedonic models for different housing type. Aggregate hedonic models are applied to the GTA for the period 1996 to 2017 for the single-family house, condominium, and townhouse markets. By using aggregate data, a spatio-temporal Durbin panel framework can be used to test its ability to represent additional insights not present in simpler models.

The objective of this investigation is to explore the edges of operationalized theory in spatial econometrics. This will form the basis for future applications to individual sales data and inform necessary adjustments to the model.

Methodology

Data Preparation

The Toronto Real Estate Board (TREB) provides monthly housing market statistics for the GTA dating from 1996 to the present (*13*). They provide sale totals, average sale prices, and median sale prices for a custom zonal system, which defines relatively small zones within the City of Toronto and larger zones for the outlying communities. Thus, the GTA is divided into a total of 65 zones. TREB zone-based data are used in this study for separate dwelling types: single-family houses, condominiums, and townhouses. A subset of the data are used, comprising of March and September of each year, as these represent months with average sales based on historical data reported by TREB (*13*).

Additional data were collected from Statistics Canada for a variety of socioeconomic metrics and reconciled to the TREB zonal system. Data from Statistics Canada includes income, employment, and population totals for each zone. In addition, to capture the patterns of urban form and transportation services in the GTA, aggregate trip-mode specific information were compiled from the National Household Survey. The distance parameter in the model is based on the distance between zone centroids and the Toronto central business district (CBD), represented by the centroid of zone C08. Colwell and Munneke examined the validity of a uniform distance gradient for a large metropolitan region (*14*). They argued that directional heterogeneities will arise because of racial segregation and the development of secondary business centres. With this in mind, additional variables were introduced to account for directional effects and variables defined to categorize zones as one of SW, NW, or NE of the CBD (no zones are SE of C08). Additional variables were generated to represent the effects of the major rail transport services in the GTA: TTC subway, TTC streetcar, and GO Transit commuter rail. Each community/zone is classified based on its adjacency to each of these rail systems.

Historical data were utilized to determine times of network expansion within the study period and adjacency modified for each zone based on these data (*15–17*). In addition, a slight adjustment is made to transit adjacency factors to account for differences in access. In the case of access to the Go Train, a simple dummy variable is deemed sufficient as there is an expectation of travel time to the station location. However, in the case of the subway and streetcar network there is a variable level of access to the system for each zone. Zones are sufficiently large within the City of Toronto (about 1 to 10 km) meaning that access quality varies. Consideration was made for inverse distance from the zone centroid to station locations, but this would not provide a meaningful accessibility measure for the streetcar network because it has a high station density. A simple model is employed in which zones with transit lines through their centre are given a value of 1.0, while those with a line along their border or terminating a short distance into the zone are given a value of 0.5. This provides a basic measure of no access, minimal access, and good access to rail transit. The use of aggregate data influences the interpretation of transit variables in the model. Seo et al. found most models either consider distance to transit lines or distance to transit nodes (i.e., stations) (*18*). These alternative interpretations can affect the sign of the parameter because proximity to a transit line may not exhibit the same benefits associated with proximity to station nodes. Those residing along the transit line will experience the noise and infrastructure barriers of transit, but not the benefit of accessibility associated to nodes. In this aggregate model, average prices are considered and it is assumed that the positive effects of rail proximity govern. Suburban zones were ultimately deemed too large in the model to specify proximity to GO stations, therefore only subway and streetcar dummy variables are included in the final model structure.

Early studies of hedonic price focused on total price, rather than normalizing for dwelling or property area (*19*). These studies found a positive curve for the variation of housing price with distance from the CBD. McMillen argued this missed the consideration that richer households tend to live in large dwellings farther from the CBD (*20*). He used the Los Angeles region as an example where single-family house prices were higher farther from the CBD. A study of non-residential prices by Gingerich et al. standardized their dataset of 1991 to 2011 sales to 2011 dollars (*21*). They further sought to standardize prices to reflect unit area. Unfortunately, these data were not available and they were forced to substitute land area. Gingerich et al. argue this has drawbacks in certain dwelling types, which tend to developer vertically (e.g., apartments or offices) (*21*). This is partially mitigated in their study by separation of land-use. We considered these factors in model development, but decided to maintain use of the unadjusted price. No adjustment is made for inflation because the model contains explanatory variables with heterogeneous rates of inflation.

By maintaining nominal price valuations, we can ascertain the effects of inflation endogenously with the model. Furthermore, we explicitly examine price effects of dwelling area because initial analysis suggested this may not be a significant factor in the GTA. The average unit area in each zone was not readily available from TREB sources. As such, an estimation was devised based on a secondary dataset developed in the course of this study. We were able to pull current real estate listings for the GTA from multiple listing service (MLS) data available online. A scraping procedure was developed, which produced a database of approximately 20,000 sale and rental listings for the month of May, 2017. These data include the geo-location of the listing, its offer price, and dwelling area in square meters, and number of bedrooms. These data are included in the models by taking averages of dwelling areas for listings geo-located to each zone, distinguishing by dwelling type. Additional totals were developed for the number of bedrooms; however, this was identified as a weaker metric as there can be wide variation in the size of bedrooms and this may affect price.

We normalize price using Box–Cox transformations defined by

$$y_i^\lambda = \frac{y_i^\lambda}{\lambda} \tag{1}$$

In which λ is determined according to the profile likelihood function. Transformation to the natural logarithm forms a special case of the Box–Cox transformation when λ equals 0. This has been found to improve model fit when compared with the use of unadjusted price (*22, 23*).

Employment is represented in a series of 6 categories by employment density in each year. These statistics were obtained from national census data for 1996, 2001, 2006, 2011, and 2016. Values for the intervening years are linear interpolations. This may miss some variation (e.g., between 2006 and 2011 because of the financial downturn), but the Box–Cox transformation is believed to minimize this effect and its impact on results should be minimal.

Spatial Econometrics

The analysis of real estate markets is an inherently spatial process as the price of a dwelling is influenced, not only by its own characteristics, but also those of surrounding dwellings and community features. As such, one should question the validity of a standard linear regression model in the context of real estate prices. Anselin identifies two components to this spatial effect: spatial dependence and spatial heterogeneity (*24*). Spatial dependence refers to the potential for interdependence between observations that are geographically close and violate the assumption of uncorrelated error terms. This follows Tobler's First Law of Geography, which states that geographically specified features will have more in common with spatially proximate features than those further away. Models that account for this dependence are termed spatial lag models. Spatial heterogeneity refers to systematic variation across space producing heteroskedastic error terms and this is described by spatial error models. Some combination of these effects exists in real estate markets and it is an objective of spatial econometrics to account for these effects. Previous research in the GTA conducted by Haider and Habib made use of these techniques to produce spatially-consistent models of price, but neither produced consistent panel models through time (*8*, *9*). Rather, each examined annual data in isolation. This investigation complements their work by further considering temporal effects within spatial hedonic price models.

The spatial lag model is defined as

$$\boldsymbol{P} = \beta X + \rho \boldsymbol{WP} + \phi + \boldsymbol{c} + \boldsymbol{\alpha} + \boldsymbol{v} \tag{2}$$

In which $\boldsymbol{P}$ is the price matrix for each of the N zones and T time periods. The spatial interaction effect is measured for $\boldsymbol{P}$ by $\boldsymbol{WP}$ with respect to each of the $\boldsymbol{j}$ adjacent zone and ρ is a fixed response parameter. β is a Kx1 vector of regression estimators, X is a 1xK vector of independent explanatory variables, W is a spatial weights matrix that is row-standardized, and ϕ is a constant intercept term. Optional parameters c and $\boldsymbol{\alpha}$ can be estimated and represent specific effects for each of the N zones and T time periods, respectively. If specific effects are treated as fixed, the intercept ϕ can only be estimated under the condition that $\sum_i c_i = 0$ and $\sum_t \alpha_i = 0$. Finally, v is an IID error term specified for each of the zone-time pair.

Similarly, the spatial error model is defined as

$$\boldsymbol{P} = \boldsymbol{\beta X} + \phi + \boldsymbol{c} + \boldsymbol{\alpha} + \boldsymbol{U} \tag{3}$$

In which parameters are defined as in the spatial lag model and $\boldsymbol{U}$ is an error term defined as

$$\boldsymbol{U} = \lambda \boldsymbol{WU} + \boldsymbol{V} \tag{4}$$

In which λ is the spatial autocorrelation coefficient and the error of each zone is assumed to be dependent on the error in adjacent zones plus a zone-specific term, $\boldsymbol{V}$.

The spatial Durbin panel model (SDPM) is a generalization of the spatial lag model, adding spatially lagged explanatory variables, and is defined as

$$\boldsymbol{P} = \boldsymbol{\beta X} + \theta \boldsymbol{WX} + \rho \boldsymbol{WY} + \phi + \boldsymbol{c} + \boldsymbol{\alpha} + \boldsymbol{V} \tag{5}$$

In which $\boldsymbol{\theta}$ is a vector of estimators similar to $\boldsymbol{\beta}$. The validity of the SDPM as an improvement over both the spatial lag and spatial error models is an important topic in the spatial econometric literature (*25*). It can be compared with the spatial lag model by testing the hypothesis $H_0 : \boldsymbol{\theta} = 0$ and the spatial error model by testing the hypothesis $H_0 : \boldsymbol{\theta} + \lambda\boldsymbol{\beta} = 0$. These can be tested according to a Wald test or, with the estimation of spatial lag and error models, a likelihood ratio (LR) test. Each test has different applications in which it is most effective: the Wald test is more sensitive to parametrization of nonlinear constraints and LR test requires additional model estimations (*25*).

We consider three forms of the SDPM based on methods outlined by Elhorst (*25*): (i) a fixed effects model, (ii) a bias-corrected robust fixed effects model, and (iii) a random effects model. By pooling data according to spatial and temporal ensembles, the model does not account for spatial or temporal heterogeneity between zones and periods. This is remedied by introducing zonal effect terms for time-invariant specific effects and time period terms for space-invariant specific effects (the c_i and α_t terms in Equations 1, 2, and 4) (*26*). This accounts for difficult to obtain or measure variables specific to each zone and time period. For example, one spatial unit may be located near the lakeshore, while another only abuts it at a corner. This may produce different prices in reality, which are challenging to represent in model specification without introduction of large numbers of explanatory variables. In a random effects model, fixed spatial effect terms are replaced by a probabilistic function. The effect of each zone on all other zones is considered individually and an assumption is made concerning the distribution (e.g., normal or gamma) (*25*, *27*). The decision to adopt random effects, rather than fixed effects, can be determined from a Hausman Test, mathematical derivation for spatial panel models is available in Lee and Yu (*28*). We use the maximum likelihood approach of estimation for the SDPM.

In the case of such a model, Lee and Yu find that a bias is introduced in the fixed effects process when the demeaned values are estimated directly (*29*). The size of the bias depends on the number of time periods T and number of spatial zones N. We employ both spatial and temporal fixed effects in our model, which are found to be biased when both T and N are large. For a discussion of the bias correction process, see Lee and Yu (*29*).

A strength of the SDPM model is its facilitation of robust direct and indirect impact analysis. Direct impacts can be defined as the effect of changing an explanatory variable on the dependent variable in its zone. For example, the impact of higher average income on the price of a single-family house. Indirect impacts measure the effect of changes in an explanatory variable in one zone on the dependent variable in all other zones. This provides a measure of spatial spill-over effects, which can be helpful in providing a broader analysis of real estate prices. The direct and indirect impacts for the SDPM are derived by Elhorst according to the partial derivative method recommended by LeSage and Pace (*25*, *27*). LeSage and Pace define a direct effect as the average of the diagonal elements of the right-hand partial derivative matrix below and the indirect effect as the average of the row or column sums across this matrix, including off-diagonal elements only (*27*). These impacts are calculated across the spatial zones for each of the k explanatory variables at a particular time t according to

$$\frac{\partial \boldsymbol{P}}{\partial \boldsymbol{X}_k} = (\boldsymbol{I} - \lambda \boldsymbol{W})^{-1}(\beta_k \boldsymbol{I} + \theta_k \boldsymbol{I}^* \boldsymbol{W}) \quad (6)$$

In which $\boldsymbol{I}^*$ is a matrix with values of zero along its diagonal and one for all off-diagonal elements. The direct and indirect impacts cannot be determined from point estimators and require methods of repeated sampling to obtain a distribution from which to test for significance. This is provided in Matlab and R software (*26*).

Analysis of Greater Toronto Area

Statistical Tests for Model Specification

We begin with, we will describe some general statistical analysis to examine regional spatial structures. Previous research suggested the GTA is a polycentric region and assumed spatial clustering around ten regional shopping centres (*8*). The reasons for re-examining this assumption are two-fold. Firstly, foundational location theory is structured around the monocentric city and assumes land prices follow a declining gradient from the CBD (*30–32*). A devision towards polycentricity would influence explanatory variables included in the model to test sensitivity of price with distance to the CBD. Secondly, recent developments in the field have produced more robust tests of spatial clustering. Arribas-Bel and Sanz-Gracia used the method of LISA (Local Indicators of Spatial Association) plots across the 359 metropolitan areas of the United States for three census years (1990, 2000, and 2010) (*33*). LISA are based on the local Moran I statistic calculated as

$$I = \frac{\frac{N}{S_0}\sum_i \sum_j W_{ij} Z_i Z_j}{\sum_i Z_i^2} \quad (7)$$

In which I is the value of the statistic, Z_i is the deviation of the variable of interest with respect to the mean, W_{ij} is the matrix of spatial weights, and $S_0 = \sum_i \sum_j W_{ij}$. We use an adjacency spatial weights matrix, based on a binary check of whether two zones share a boundary line. This statistic can be calculated for each zone and provides a measure of its deviation from the mean. The significance of each parameter can then be defined by running a randomized assignment of values to each zone. If the deviation obtained using the Local Moran I statistic is greater than that obtained through random variation, it is said to be significant. A LISA plot can be developed, which classifies each zone as to whether it has a significantly high value of the variable of interest and if its neighbours also have high values. Similar indicators can be developed for High-Low, Low-High, and Low-Low for any variable of interest. Arribas-Bel and Sanz-Gracia state that whether a region exhibits polycentricity is partially a function of the quality of its transportation network (*33*). Polycentricity will evolve from high generalized costs of travel, which shorten the average trip length. This will lead to localized clusters of employment throughout the region. In contrast, monocentric regions most likely have strong transportation network, especially fixed rail corridors, which minimize the generalized cost of travel. They found monocentric city structures remained dominant in the US in 2010 at 57.7% of metropolitan areas, with no significant difference from the 56.5% of 20 years prior.

In this analysis we employed the methods outlined by Arribas-Bel and Sanz-Gracia to examine clustering of employment, income, and single-family house price (*33*). This is illustrated in Figures 1, 2 and 3 for 2016.

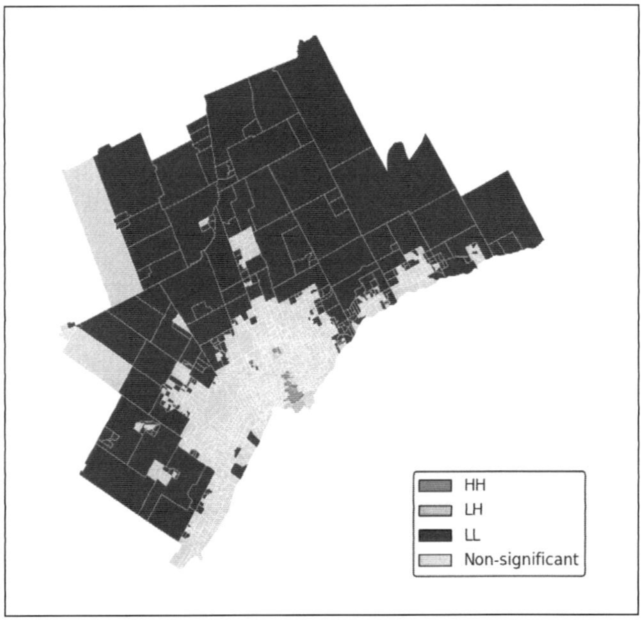

Figure 1. Employment density defined by TREB zones for 2016.

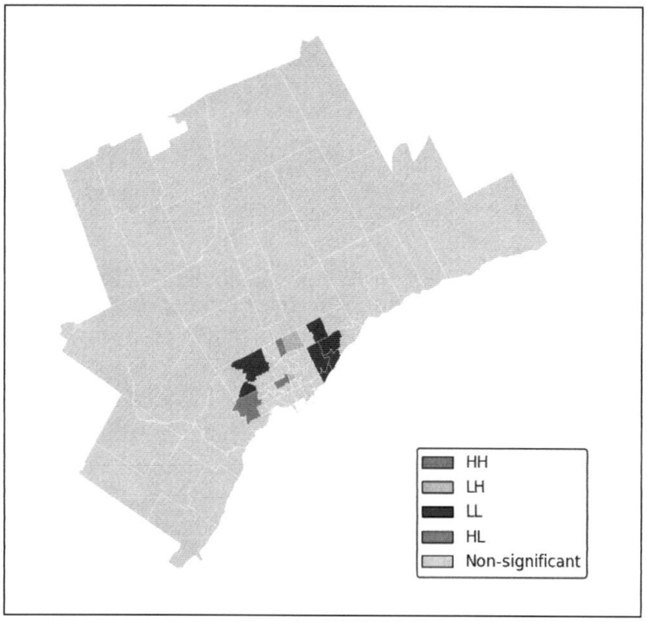

Figure 2. Average income defined by TREB zones.

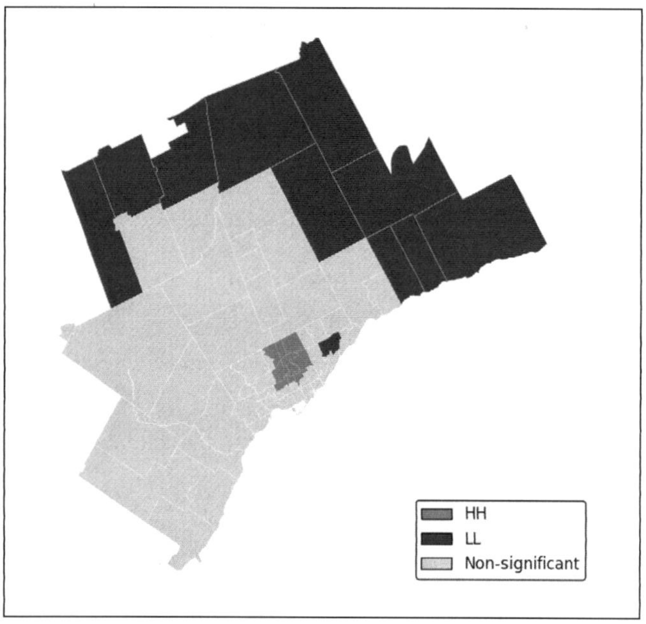

Figure 3. Single-family house price defined by TREB zones.

Employment density is used, rather than total employment, in Figure 1 because zone areas are not uniform. This indicates a spatial cluster of high employment density in the CBD and a minor cluster to its north. It can also be seen that there is very little employment in the outlying zones, but a region of uniform employment density exists within the City of Toronto outside the CBD, which extends into Mississauga to the west. This supports the notion of employment diffusing across the

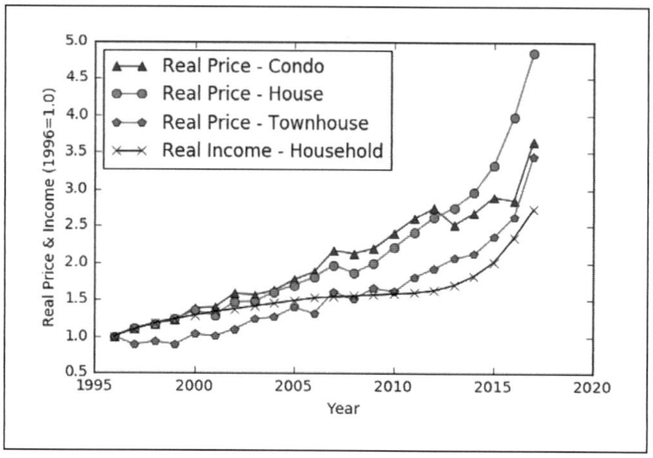

Figure 4. Change in real price of dwellings compared with real income of households (1996 = 1.0).

region, rather than becoming concentrated in a set of suburban centres. This does not support earlier findings of polycentricity by Haider (*8*). Additional analysis was conducted with greater geographic detail in regions outside the City of Toronto using census dissemination areas, which further confirmed the monocentricity of the GTA, with indications of a minor employment centre in Mississauga. However, these results were not included in this publication for brevity and to maintain consistent zone definitions throughout the text.

The LISA plots for income and single-family house price largely conform with the pattern observed in employment density. They also confirm that earlier assumptions of polycentricity appear to be erroneous according to standard methods of spatial cluster analysis (*8*). An additional region of low values was observed for income and housing price along the eastern periphery of the City of Toronto in Scarborough. This fits with the expectation that the spatial location of low income households should correspond with regions of low single-family house prices, regardless of employment density.

An illuminating plot developed during data preparation provides comparison of the real price for each dwelling type with real household income (Figure 4). This contrasts the purchasing power of GTA residents against rising housing price, with each being standardized against its magnitude in 1996. There are certainly caveats associated with using average household income as an indicator of purchasing power as we must neglect the effect of rising prices in other commodity markets on real income. However, it provides a strong illustration of changing conditions over the study period. It can be seen that single-family house prices rose relative to real income beginning in 2007–2008 and roughly doubled their level from 1996 compared to the purchasing power of

Table 1. Summary of Model Specifications

	Condominiums			Houses			Townhouses		
Test statistic	Fixed	Fixed-robust	Random	Fixed	Fixed-robust	Random	Fixed	Fixed-robust	Random
Log likelihood	−8586.1	−8586.1	-3.8×10^9	−10148.5	−10148.5	-1.6×10^9	−9014.6	−9014.6	-4.9×10^7
(pseudo)R^2	0.7974	0.7973	0.7619	0.9389	0.9389	0.9228	0.7182	0.7180	0.6489
corr-(pseudo)R^2	0.1772	0.1774	0.5108	0.4170	0.4171	0.7704	0.0899	0.0900	0.4504
σ^2	27.27	28.36	28.27	83.43	86.73	85.79	37.06	38.54	38.14
Wald test lag	112.13*	110.20*	78.98*	180.68*	170.27*	167.96*	50.20*	49.09*	35.89*
LR test lag	110.55*	110.55*	–	174.10*	174.10*	–	49.84*	49.84*	–
Wald test error	115.69*	109.37*	80.71*	180.56*	169.87*	169.69*	51.03*	47.88*	36.45*
LR test error	112.51*	112.51*	–	175.37*	175.37*	–	50.45*	50.45*	–
Hausman test	39.84*			6.52			14.54		

Note: LR = likelihood ratio. *95% CI.

households. Conversely, condominium and townhouse prices have not risen as precipitously, supporting the notion that households may choose to change their residence consumption patterns in the coming years. Households in the GTA could obtain additional utility from increased real income by re-locating into condominiums or townhouses rather than allocating an increasingly larger share of income to consumption of single-family houses.

Standard linear regression models were developed for each of the dwelling types to illustrate model improvement with the inclusion of each feature of the final model. The unadjusted average price is used in this initial model and the pseudo-R^2 used as a measure of fit. In the case of both condominiums and townhouses, the pseudo-R^2 is in the range of 0.33 to 0.34. Single-family houses show a stronger fit with this simple specification, with a pseudo-R^2 of 0.71. Significant improvements are obtained in all cases with the introduction of Box–Cox transformations of the price vector. This produces pseudo-R^2 values of 0.64, 0.85, and 0.58 for each of condominiums, houses, and townhouses, respectively. Table 1 outlines the model specifications for the SDPM tested in this study for each of the dwelling types. Model 1 and 2 represent specifications with fixed effects for space and time, with Model 2 differing only in its use of the bias-correction proposed by Lee and Yu (*29*). Model 3 employs random effects across the spatial dimension of 65 zones and fixed effects across the 43 time periods. Each model specification provides a strong fit according to the pseduo-R^2 and all models are significant improvements over both the spatial lag and spatial error specifications according to the Wald test. The LR test confirms these results for Models 1 and 2, but consistent results could not be obtained for the random effects assumption, Model 3. The log-likelihood is several orders of magnitude lower for Model 3 with respect to the fixed effects models, but the pseudo-R^2 is not significantly lower. In addition, a corrected pseudo-R^2 that removes spatial and temporal effect terms finds Model 3 to provide a much stronger fit based solely on the explanatory variables. This is borne out by the results of the Hausman test, which recommends use of random effects over fixed effects for both houses and townhouses. Fixed effects are assumed for condominiums and a summary of parameter values is provided for each model in Table 2.

Spatial Durbin Panel Model of GTA

We begin the summary of model results with a discussion of geographic considerations. Analysis discussed above found the GTA to be a monocentric region with a strong influence by the CBD. This validates the inclusion of distance to the CBD as an important explanatory variable in the model. This was confirmed by this parameter being statistically significant for each of dwelling types: condominiums, houses, and townhouses. The direction from the CBD was also compared for each dwelling type and estimators found to be significant; however, there does not appear to be a significant price heterogeneity with direction when relative parameter magnitudes are compared between directional variables. Factors associated with market dynamics were explored via the inclusion of a dummy variable for the month of sale, mortgage interest rate at time of sale, and total sales in the zone. Only total sales were left in the model as the other two variables produced singular matrices, which could not be solved. Total sales were not found to be significant, nor were the number of sales in adjacent zones.

Average income was found to be significant for each of the dwelling types and the parameter for houses to be significantly higher than could be accounted for based solely on higher price levels in this market relative to condominiums and townhouses. This suggested average income plays a stronger role in determining single-family house prices than either condominiums or townhouses. Interestingly, the presence of high incomes in adjacent zones tended to suggest lower prices in a zone for

Table 2. Summary of Model Results

Parameter	Condominium (fixed-robust)	House (random)	Townhouse (random)
Distance to CBD (km)	–0.047*	–0.145*	–0.059*
SW of CBD dummy	14.769*	34.013*	15.787*
NW of CBD dummy	14.754*	32.562*	16.396*
NE of CBD dummy	16.838*	32.099*	15.873*
Total sales	–0.008	–0.025*	0.012
Average income ($10,000)	0.887*	4.146*	0.592*
Proportion trips by active mode	9.653	49.930*	34.487*
Average dwelling size (m^2)	0.0004	–0.019*	–0.008
Employment density (ordinal)	2.268*	5.878*	2.366*
Population density (100s persons/km^2)	0.062*	0.266*	0.057*
Subway dummy	5.163*	9.479*	1.411
Streetcar dummy	13.502*	3.095	–3.259
W*Sales	0.017	0.0006	–0.030
W*AveIncome	–1.050*	0.968*	0.188
W*PropActive	–14.377	–186.472*	–49.504*
W*SQM	–0.011	0.032	–0.031
W*EmpCode	0.091	–3.675	–1.309
W*PopDens	0.180*	0.615*	0.156*
W*SubwayDummy	–30.072*	16.961	–0.920
W*StreetcarDummy	–15.431*	1.042	1.488
Spatial autocorrelation (ρ)	0.013	–0.0009	–0.021
θ	NA	0.111*	0.198*

Note: CBD = central business district; W = contiguity weight matrix. *95% CI.

condominiums, but higher prices for houses. This may be because of high income communities dominated by single-family houses being clustered, while there is more dispersion in the income distribution of condominium residents. These result are strengthened by access to the direct and indirect effect vectors. From this analysis we found the spill-over effects from average income to be positive for single-family houses, but negative for condominiums. Population and employment density were both found to have significant positive effects on prices in all markets.

In relation to transportation indicators, the presence of a streetcar line tended to have a strong influence on condominium prices, while the subway appeared significant for condominiums and houses. The use of a Durbin specification provided an indication that, not only did the presence of a streetcar line seem to explain higher prices, its presence in adjacent zones was associated with lower condominium prices. This effect was also observed for subway lines, which is indicative of a mirroring effect in the parameters. That is, the zone system results in zones either: (1) containing a transit line or (2) not containing a transit line, but the adjacent zone containing one. This produced a positive effect on condominium prices for zones indicated to have a transit line and a mirrored negative effect from their corresponding spatially weighted variables for zones lacking rail transit. The proportion of trips to work by active mode was a significant factor. This variable acted as a surrogate for walk access in the zones and was found to be an important factor in determining price levels. The model suggested this variable is significant and positive for house and townhouse prices and exerts a significant negative influence on adjacent zones. Figure 3 supports this result, with housing prices in the more walkable City of Toronto being much higher than those in outlying municipalities. There are certainly many other factors contributing towards this difference, but this provides an initial indication of the desirability of an urban spatial configuration.

Spatial and temporal fixed effects were compared for each dwelling type according to the robust fixed effects Model 2. Random effects are more challenging to compare as they comprise $N \times N$ and $T \times T$ matrices for each zone and period, respectively. Despite providing a large portion of the explanation for price levels, very few of the fixed effects were found to be individually significant at the 95% level. No temporal effects were deemed significant, but parameter magnitudes tended to increase over time in line with inflation of price. Each of the condominium and house markets has six and five, respectively, zones with significant effects on price, but these are not common between dwelling types. No spatial effects were of note from the townhouse model. In the condominium and house models, location in CBD zone C08 had a positive influence on price, while location in zone E03 had a negative influence in both markets. High prices in the CBD were expected, while a preliminary study of Toronto crime maps suggested the lower prices in zone E03 may be because of elevated rates of major crime in the area (*34*). This analysis suggested simple models could be run with fixed effects and variation in parameter estimates explored to determine factors in each zone and time period producing heterogeneity. Explanatory variables could then be based on these factors. LeSage and Pace suggest that spatial lags may not be necessary in cases where fixed zonal effects dominate (*27*). It was determined that spatial lag effects are significant in the present set of models and their inclusion is justified based on improvement of the log-likelihood.

Conclusions and Future Research

This study utilized advanced methods of spatial econometrics to develop a series of hedonic price models for the highly dynamic and volatile Toronto residential property market. Average income, distance to the CBD, and population and employment densities were found to be significant factors in the determination of prices. The model found spill-over effects to be important in the property market, with average income in adjacent areas having a positive influence on house and a negative effect on condominium prices. Spatial and temporal effects were compared between fixed and random specifications. It was clear that fixed systematic effects dominate in the condominium market, while these effects tend towards randomness in the house and townhouse markets. Comparisons were made with previous hedonic residential price models developed for the same study area and it was identified that previous assumptions of polycentricity of employment do not hold when subjected to rigorous analysis. The present model formulation was able to accomplish similar levels of model fit, with relatively few explanatory variables with respect to the previous models developed for the same region. This suggests a robust representation of the market can be obtained without the need to obtain large numbers of variables, many of which can be dubious in their validity. With an extension to individual sales records, a single hedonic price model could be developed that includes structure type-specific variables. The presence of an apartment building near a single-family dwelling can affect its price.

This study forms a strong foundation for residential hedonic models of the GTA. However, it is constrained by the aggregate nature of current SDPM and would benefit from analysis of individual listings. It may be possible to devise a composite model, which would use individual sales data as the dependent variable and maintain the attractive features of the SDPM specification. Fixed effects would remain defined at the level of zones to minimize computation time, but dwelling-specific variables could be added. This investigation suggested there are benefits to households in a switch away from single-family houses. A more explicit study of this shift would be useful, with consideration of transportation accessibility and real income gains.

Author Contributions

The authors confirm contribution to the paper as follows: study conception and design: Jason Hawkins, Khandker Nurul Habib; data collection: Jason Hawkins; analysis and interpretation of results: Jason Hawkins; draft manuscript preparation: Jason Hawkins and Khandker Nurul Habib. All authors reviewed the results and approved the final version of the manuscript.

References

1. Bonnell, G. Toronto Housing Market at 'Turning Point,' Realtor Says. BNN Bloomberg, June 2017. https://www.bnnbloomberg.ca/toronto-housing-market-at-a-turning-point-realtor-says-1.768854. Accessed June 14, 2017.
2. Burda, C. *Location Matters Series: Understanding the Factors Affecting Home Prices in the GTA*. RBC-Pembina, Toronto, 2013.
3. Concas, S. Accessibility and Housing Price Resilience: Evidence from Limited-access Roadways in Florida. *Transportation Research Record: Journal of the Transportation Research Board*, 2013. 2357: 66–76.
4. Diao, M., and J. J. Ferreira. Residential Property Values and the Built Environment: Empirical study in the Boston, Massachusetts, Metropolitan Area. *Transportation Research Record: Journal of the Transportation Research Board*, 2010. 2174: 138–147.
5. Dorantes, L. M., A. Paez, and J. M. Vassallo. Analysis of House Prices to Assess Economic Impacts of New Public Transport Infrastructure Madrid Metro Line 12. *Transportation Research Record: Journal of the Transportation Research Board*, 2011. 2245: 131–139.
6. Shin, K., S. Washington, and K. Choi. Effects of Transportation Accessibility on Residential Property Values: Application of Spatial Hedonic Price Model in Seoul, South Korea, Metropolitan Area. *Transportation Research Record: Journal of the Transportation Research Board*, 2007. 1994: 66–73.
7. MetroLinx. *The Big Move: Transforming Transportation in the Greater Toronto and Hamilton Area*. Greater Toronto Transportation Authority, 2008.
8. Haider, M. *Development of Hedonic Prices Indices for Freehold Properties in the Greater Toronto Area: Application of Spatial Autoregressive Techniques*. PhD thesis. University of Toronto, 1999.
9. Habib, M. A. *Microsimulating Residential Mobility and Location Choice Processes within an Integrated Land Use and Transportation Modelling System*. PhD thesis. University of Toronto, 2009.
10. Pace, R. K., R. Barry, O. W. Gilley, C. F. Sirmans, and E. J. Ourso. A Method for Spatial-Temporal Forecasting with an Application to Real Estate Prices. *International Journal of Forecasting*, Vol. 16, 2000, pp. 229–246.
11. Otto, P., and W. Schmid. Spatiotemporal Analysis of German Real-Estate Prices. *The Annals of Regional Science*, Vol. 60, No. 1, 2016, pp. 41–72.

12. Iacono, M., and D. Levinson. Location, Regional Accessibility, and Price Effects: Evidence from Home Sales in Hennepin County, Minnesota. *Transportation Research Record: Journal of the Transportation Research Board*, 2011. 2245: 87–94.
13. Toronto Real Estate Board. *Market Watch*, TREB, Toronto, 2017.
14. Colwell, P. F., and H. J. Munneke. Directional Land Price Gradients. *Journal of Real Estate Financial Economics*, Vol. 39, No. 1, 2009, pp. 1–23.
15. Toronto Transit Commission. *TTC Milestones*, TTC, Toronto, 2013.
16. Parrish, B. *Eves Government Opens New GO Transit Station at York University*, Ontario Ministry of Transportation, Ottawa, 2003.
17. Garcia, D., and J. Bow. *Go Transit's Barrie Line*, *Transit Toronto*, 2017. http://transit.toronto.on.ca/regional/2105.shtml. Accessed June 15, 2017.
18. Seo, K., A. Golub, and M. Kuby. Combined Impacts of Highways and Light Rail Transit on Residential Property Values: A Spatial Hedonic Price Model for Phoenix, Arizona. *Journal of Transport Geography*, Vol. 41, 2014, pp. 53–62.
19. Abelson, P., R. Joyeux, and S. Mahuteau. Modelling House Prices Across Sydney. *Australian Economic Review*, Vol. 46, No. 3, 2013, pp. 269–285.
20. McMillen, D. P. The Return of Centralization to Chicago: Using Repeat Sales to Identify Changes in House Price Distance Gradients. *Regional Science and Urban Economics*, Vol. 33, No. 3, 2003, pp. 287–304.
21. Gingerich, K., H. Maoh, K. Gingerich, H. Maoh, and W. Anderson. Location and Transportation Effects on Nonresidential Real Estate Price Regressions in Windsor, Ontario, 39 Canada. *Transportation Research Record: Journal of the Transportation Research Board*, 2013. 2397: 99–107.
22. Dubé, J., M. Thériault, and F. Des Rosiers. Commuter Rail Accessibility and House Values: The Case of the Montreal South Shore, Canada, 1992-2009. *Transportation Research Part A: Policy and Practice*, Vol. 54, 2013, pp. 49–66.
23. Smith, J. J., T. A. Gihring, and T. Litman. Financing Transit Systems through Value Capture: An Annotated Bibliography. *American Journal of Economics and Sociology*, Vol. 65, No. 3, 2006, pp. 751–786.
24. Anselin, L. *Spatial Econometrics: Methods and Models*. Springer Netherlands, Dordrecht, 1988.
25. Elhorst, J. P. Matlab Software for Spatial Panels. *International Regional Science Review*, Vol. 37, No. 3, 2014, pp. 389–405.
26. Elhorst, J. P. Spatial Panel Models. In *Handbook of Applied Spatial Analysis* (Fischer, M., and A. Getis, eds.), Springer, Berlin Heidelberg, New York, 2010.
27. LeSage, J., and K. R. Pace. *Introduction to Spatial Econometrics*. CRC Press, New York, 2009.
28. Lee, L. F., and J. Yu. Some Recent Developments in Spatial Panel Data Models. *Regional Science and Urban Economics*, Vol. 40, No. 5, 2010, pp. 255–271.
29. Lee, L. F., and J. Yu. Estimation of Spatial Autoregressive Panel Data Models with Fixed Effects. *Journal of Econometrics*, Vol. 154, 2010, pp. 165–185.
30. Alonso, W. *Location and Land Use: Toward a General Theory of Land Rent*. Harvard University Press, Cambridge, 1964.
31. Muth, R. F. *Cities and Housing: The Spatial Pattern of Urban Residential Land Use*. University of Chicago Press, 1969.
32. Mills, E. An Aggregative Model of Resource Allocation in a Metropolitan Area. *American Economic Review*, Vol. 62, No. 2, 1967, pp. 197–210.
33. Arribas-Bel, D., and F. Sanz-Gracia. The Validity of the Monocentric City Model in a Polycentric Age: US Metropolitan Areas in 1990, 2000 and 2010. *Urban Geography*, Vol. 35, No. 7, 2014, pp. 980–997.
34. Toronto Police Service. *Major Crime Indicators*, *Toronto Police Service*, 2017. http://data.torontopolice.on.ca. Accessed July 4, 2017.

The Standing Committee on Transportation and Land Development (ADD30) peer-reviewed this paper (18-00483).

Article

TRR
JOURNAL OF THE TRANSPORTATION RESEARCH BOARD

Transportation Research Record
2018, Vol. 2672(6) 31–39

DOI: 10.1177/0361198118774159
journals.sagepub.com/home/trr
SAGE

Travel Behavior in TODs vs. Non-TODs: Using Cluster Analysis and Propensity Score Matching

Keunhyun Park[1], Reid Ewing[1], Brenda Case Scheer[1], and Shabnam Sifat Ara Khan[2]

Abstract

As a compact, mixed-use, and walkable district near a high-quality transit station, transit-oriented development (TOD) has arisen as a sustainable form of urbanism to minimize automobile dependency and maximize ridership. Existing travel behavior studies in the context of TOD, however, are limited in terms of small sample size, lack of consistency in TOD classification, and failure to control for residential self-selection. This study examines various travel outcomes—vehicle miles traveled (VMT), auto trips, transit trips, and walk trips—in different types of station areas in eight U.S. metropolitan areas using cluster analysis and propensity score matching. Using cluster analysis with three built environment factors—activity density, land use diversity, and street network design (i.e., D variables)—this study classifies existing 549 station areas as TOD, transit-adjacent development (TAD), and Hybrid types. After controlling for residential self-selection, the result shows that a TOD motivates its residents to walk more and take transit more while driving less. The significant difference between TOD and TAD in both VMT and the number of automobile trips means that TOD makes the personal vehicle trips shorter (39% reduction) and fewer (35% reduction). Travel behaviors in the Hybrid type are also examined for the potential outcomes of gradual and practical changes.

Expenditures on transportation have increased from the sixth largest share (less than 2%) of household budgets in 1917 to the second largest share since the 1970s (17% in 2014) (*1*). Under this circumstance, transit-oriented development (TOD) has gained popularity worldwide as a sustainable form of urbanism by concentrating developments near a transit station to minimize automobile dependency and maximize ridership. A TOD project should give people alternative transportation options and, in turn, decrease their transportation cost.

Much of the literature verifies that TODs enhance the use of public transport and reduce car usage (*2–7*). Existing TOD studies, however, have limits in terms of (1) small numbers of study sites, (2) lack of systematic methodology to distinguish TOD from other types of station areas, and (3) lack of control for the impact of residential self-selection on travel behavior. There are some exceptions to the above limitations (*8*), but no study overcomes all three limitations. As a result, it is hard to generalize the findings from the literature to planning practice. Also, the current distinctions between TODs and transit-adjacent developments (TADs), an opposite concept of TOD, limit the practical implications for transit officials and planners. Finally, when it fails to control self-selection effect, the result might overestimate the impact of TOD urban form on travel behavior.

Thus, this study asks two research questions. First, how can station area types be distinguished? Second, how do travel behaviors vary among different station area types? To answer these questions, cluster analysis is used to classify station area types and propensity score matching (PSM) is used to control for residential self-selection. For better generalizability, the data are collected from household travel surveys in eight urban areas across the U.S. with exact XY coordinates for households and trip ends. For greater policy relevance, the data are used to analyze various travel outcomes such as an automobile, transit, and walk trips at household level for different station area types. By doing so, this study seeks to examine the pure impact of living in TOD—versus a TAD—on travel behavior.

There is broad interest in the planning and policy communities for accurate tools to predict the consequences of TOD on the generation of transit ridership and reduction of automobile usage. This analysis will help guide transportation planners and decision makers to evaluate TOD projects regarding their economic, social, and environmental impacts.

[1]Department of City & Metropolitan Planning, University of Utah, Salt Lake City, UT
[2]City of Spearfish, South Dakota, Spearfish, SD

Corresponding Author:
Address correspondence to Keunhyun Park: keunhyun.park1@gmail.com

Literature Review

TOD/TAD Classification

Bernick and Cervero define TOD as "a compact, mixed-use community, centered around a transit station that, by design, invites residents, workers, and shoppers to drive their cars less and ride mass transit more" (*9*, p. 5). Kamruzzaman et al. state that a TOD is a neighborhood that is served by public transit services and offers amenities such as density, walkable, well-connected street patterns, and diversified land uses (*10*). TAD is often defined as a failure of a TOD. A TAD is a noncompact, segregated neighborhood development that calls for automobile use instead of inviting walk trips (*11–13*).

The most frequently studied factors for distinguishing a TOD from other types of station areas have been residential and employment density (*10, 14–20*), land use diversity (*10, 14, 17, 19–21*), and street connectivity (*14–16, 21–25*). Recent studies deal with all three factors in the analysis (*10, 14, 17*). There are several ways to distinguish TOD from TAD, such as cluster analysis (*10, 21*) or scoring system (*14–17*).

Existing studies differentiating TOD from TAD are limited. First, most studies cover only a single or few regions. Although Renne and Ewing study 54 regions across the U.S., the outcome variable is not overall travel behavior, but only the percentage of people who commute via public transportation (*14*). In contrast, the present study includes eight metropolitan areas in the U.S. with varying geographic and socioeconomic conditions to examine various travel outcomes. Second, unlike existing studies relying on straight-line catchment areas (*21*) or simple scoring systems (*14*), this study uses network distance from each station and cluster analysis. Finally, while Kamruzzaman et al. use a robust method of classification, their study analyzes all neighborhoods in a single city, Brisbane, Australia (*10*). Instead, this study uses the station-based approach as a focus of TOD and TAD because it deals with built environments of station areas and their impact on travel behavior, which has more direct implications for planning practice.

TOD and Travel Outcomes

Potential benefits of TOD are multiple, from promoting active modes of transportation to improving access to opportunities such as jobs or entertainment, to offering alternative mobility options and affordable housing, to reducing greenhouse gas emissions (*26, 27*). Thus, TOD serves interrelated goals of making communities socially, economically, and environmentally more robust and sustainable.

To achieve these multiple goals, a TOD should first create settings that prompt people to drive less and ride public transit more (*3*). The Center for Transit Oriented Development identifies vehicle miles traveled (VMT) as the key performance measure for TOD (*28*). Lower VMT means that people walk, bike, and use transit more, and have more transportation options.

Much of the literature verifies that TODs enhance the use of public transport and reduce car usage (*2–7*). Based on data from 17 TOD projects, Cervero and Arrington show that residents living in TOD areas are two to five times more likely to commute by transit than their non-TOD counterparts (*29*). Nasri and Zhang find that people living in TOD areas tend to drive less, reducing their VMT by around 21–38%, compared with the residents of the non-TOD areas (*5*).

Cervero (*3*) finds evidence that many TOD ridership gains are a result of self-selection—individuals who wish to drive less may select transit-oriented environments. Many studies have found associations between attitudes and travel choices as evidence of residential self-selection (*30–32*). Thus, individuals' attitudes may confound the relationship between the TOD-type urban form and travel choices and, in turn, the effect of the built environment on travel may be overestimated (*33*).

From a review of 38 empirical studies, Cao et al. examine different methodological solutions to self-selection bias (*30*). Among the methodologies, the PSM method is highly recommended in a non-randomized observational study (*30*). The propensity score approach has recently been applied in travel behavior research (*34–38*), but rarely used in the context of station areas. The authors found only one master's thesis using PSM to explore the impact of TOD on travel mode choice (*39*). A detailed explanation of the PSM is presented in the following Research Design section.

Research Design

Study Regions

This study includes eight U.S. metropolitan regions meeting three criteria. First, they must have household travel survey data with XY coordinates for households and trip ends. Second, a region must provide land use databases at the parcel level so that land use mix can be calculated for the same years as the household travel surveys. Third, they must have had a rail-based transit system before the survey was conducted.

In the eight regions (Table 1), the household travel survey was conducted between 2006 and 2012, and there were 549 rail-based transit stations on the national TOD Database (Center for Transit Oriented Development, http://toddata.cnt.org/). Transit types include heavy rail (109 stations), commuter rail (148 stations), and light rail (272 stations). Boston has the greatest number of stations (n = 239), followed by Portland (n = 94) and Miami (n = 50), and Minneapolis-St. Paul has the least number (n = 20).

Although they were conducted by individual regions' organizations such as metropolitan planning organization (MPO), the regional household travel surveys have quite similar structure and questions, akin to U.S. DOT's National Household Travel Survey (NHTS). To gather comprehensive data on travel and transportation patterns, the survey data consistently include, but are not limited to, household demographic information, vehicle information, and data about one-way trips taken during a designated 24-hour period,

Table 1. Study Regions and Transit Stations[a]

No.	Region	Year (survey)	Heavy rail	Commuter rail	Light rail	Total	Household (½ mile)
1	Atlanta, GA	2011	38	0	0	38	138
2	Boston, MA	2011	49	121	72	239[b]	1586
3	Denver, CO	2010	0	0	36	36	152
4	Miami, FL	2009	22	4	24[c]	50	26
5	Minneapolis-St. Paul, MN	2010	0	4	16	20	97
6	Portland, OR	2011	0	7	87	94	304
7	Salt Lake City, UT	2012	0	1	36	37	114
8	Seattle, WA	2006	0	11	25	35[b]	16
	Total		109	148	272	549	2433

[a]This study includes only transit stations which had opened before the survey year.
[b]The total number of stations is not equal to the sum of the columns because some stations have two or more types of transit systems.
[c]Miami's People Mover, an automated guideway transit, is included under the LRT category.

including travel time, mode of transportation, and purpose of the trip. The survey data have exact XY coordinates, enabling geocoding of the precise locations of households and measurement of the lengths of trips, whereas the NHTS provides the location of households only at the Census Tract level. The regional survey data were acquired from individual MPOs or state DOTs with confidentiality agreements.

Data

Following the definition of TOD and the literature review, this study includes "activity density," "land use diversity," and "street network design" to classify station area types. For the density variable, population and employment data for traffic analysis zones (TAZ) were acquired from regional MPOs and summed to compute an overall activity density. Activity density is the sum of population and employment within the station area, divided by gross land area in square mile (*40*). For the diversity variable, an entropy index was computed.[1] Each region provided parcel maps, enabling calculation of the proportion of the area of each land use type—residential, commercial, and public—in a ½ mile (about 800 m) buffer from each station. For the street network design variable, the number of intersections per square mile was computed from street network shapefiles. Because these three built environment variables—activity density, land use entropy, and intersection density—vary in range, each variable was standardized to a mean of 0 and a standard deviation of 1.

In addition, a "distance to transit" variable was measured as a network distance from a household to the closest rail station, because that might be an important determinant of transit trips. Also, regional accessibility is another important variable to predict travel behaviors (*40*). That variable is defined as the percentage of jobs that can be reached within 30 minutes by transit, which tends to be highest at central locations and lowest at peripheral ones. This study used travel time skims and TAZ-level employment data acquired from regional MPOs.

Station areas were drawn as a ½-mile buffer in network distance from each station. Then, individual households were allotted to their nearest station based on network distance. The resulting pooled data set in station areas consists of 2,433 households in the eight regions. From the travel survey data in eight regions, VMT, automobile trips, transit trips, and walk trips by individual households were calculated. The survey data also include demographic variables such as household size, the number of employed, and household income.

Research Process and Methods

Step 1. TOD/TAD Classification: Cluster Analysis. Because the built environments around transit stations fall within a TOD–TAD spectrum, not a simple dichotomous scale, and there is no certain agreement of ideal built environments for TOD, identifying TODs and distinguishing them from TADs could be a difficult but important research step. Cluster analysis has been a preferred method for generating TOD typologies in previous studies (*21*, *23*, *41*).

Using cluster analysis, this study classifies station area types based on three built environment factors: activity density, land use diversity, and street network design. This approach enables grouping of existing station areas based on their actual built environment characteristics, rather than theoretical criteria of TOD or TAD. To be specific, this study uses a hierarchical clustering algorithm with Ward D2 distance measure. To determine the optimal number of clusters in a dataset, this study uses the "NbClust" package in R 3.3.1 software, which provides 26 validation indices of clustering such as Calinski and Harabasz index and Silhouette index (*42*).

Step 2. Household Sample Selection: Propensity Score Matching. PSM has been widely used to overcome nonrandom assignment of treatment in the evaluation of social programs (*43*). Evaluation studies are often based on observational data, in

Table 2. Cluster Analysis Result and Descriptive Statistics

Cluster type	Number of stations	Activity density (/sq.mi.)		Entropy index		Intersection density (/sq.mi.)	
		Mean	S.D.	Mean	S.D.	Mean	S.D.
TAD	107	10,319	11,751	0.30	0.19	110	58
Hybrid	382	21,210	19,764	0.75	0.15	194	79
TOD	60	135,327	51,025	0.70	0.24	386	110
Total	549	31,559	43,821	0.66	0.24	199	108

Note: TAD = transit-adjacent development; TOD = transit-oriented development.

which the assignment of treatment is not random. Accordingly, individuals in the treatment group are likely to differ systematically from those in the control group. For example, households living in suburban regions could be more affluent than their counterparts in downtown, a result of residential self-selection. Therefore, the observed difference in behavioral outcomes between the groups is confounded by residential self-selection. Statistically, this generates a biased estimate of treatment effect.

The propensity score is defined as the conditional probability of assignment to a particular treatment given a vector of observed covariates (*44*). In the context of TOD and TAD, the treated group is households living in TOD station areas, whereas the control group is those living in either TAD or Hybrid areas.

The PSM was implemented in R 3.3.1 using *MatchIt* package. First, a binary logit model was developed to estimate propensity score using the subsample of households living in TOD (treatment) and TAD (control). Household characteristics were chosen as independent variables—household size, the number of workers, household income, distance to the nearest transit station, regional job accessibility, and the regions—as potential sources of residential self-selection and confounding factors in travel outcome. Second, each household living in TOD was matched with those in TAD based on the propensity score. Caliper length of 0.03 is used for matching, meaning that for a treatment observation, a match in control observations whose propensity scores are within 0.03 of the score of the treatment observation was searched (*45*). Third, whether the matched residents in TOD were systematically different from those in TAD was evaluated. A *t*-test was used to assess whether demographics and locational factors are balanced between the matched groups.

The final goal of PSM is to compute the true impact of TOD/TAD on travel behavior. Once the matching was complete, the average treatment effects (ATE) of station area type on VMT, transit trips, and walk trips were calculated. The ATE is computed as the mean travel factors of the matched TOD households minus those of the matched TAD households.

Because this study has three area types—TOD, TAD, and Hybrid, in contrast to most studies with only one treatment and one control—this study applies PSM with multiple nominal treatments (*46*). In this case, the propensity score is estimated separately for each pair of control and treatment. For each pair of groups, a binary logit model was estimated using the subsample of respondents living in the groups.

TOD/TAD Classification

Table 2 shows the result of hierarchical clustering. Using the *NbClust* package in R 3.3.1 software, thirteen of the 26 validation indices suggest that three is the optional number of clusters. The first cluster (n = 107) is labeled as "TAD" because it has the lowest level of density, diversity, and intersection density. The second and largest one (n = 382) is classified as "Hybrid", which has a low level of activity density and intersection density, but highest entropy index. The final cluster (n = 60) is named as "TOD" in terms that it has highest activity density and intersection density, and high level of land use mix level.

Sample households were selected as those living within half-mile network distance from stations. Individual households were allotted to their nearest stations based on network distance in order to assign the station types. TAD type has 251 households, and TOD and Hybrid type have 306 and 1,876 households, respectively (Table 3).

Table 3 shows that households living in TADs have more household members, more workers, and higher incomes than those living in TODs or Hybrids. ANOVA shows that the differences are significant. Regarding travel behavior, TAD households have much higher VMT and auto trips and lower transit and walk trips than those in TODs and Hybrids. The hybrid type is in the middle, except for their lowest household incomes and the highest level of transit trips on average.

Household Sample Selection: Propensity Score Matching

The cluster analysis results show that households living in TAD tend to be more affluent, have more cars, live in a larger household, and be more auto-oriented than their counterparts in TOD. Residential self-selection theory says, however, that the households living in TAD might live there because they are auto-oriented. Therefore, the true difference in travel outcomes between TOD and TAD is estimated here by matching samples using PSM.

Table 3. Household Characteristics and Travel Behavior by Station Area Types: Average and ANOVA Analysis

Cluster type	No. of stations	HH samples	HH size	HH workers	HH income ($1000)	VMT	Auto trips	Transit trips	Walk trips
TAD	107	251	2.19	1.28	83.72	21.23	6.06	0.72	1.91
Hybrid	382	1,876	2.15	1.22	77.02	15.44	4.93	1.47	3.89
TOD	60	306	1.54	0.97	82.02	8.61	2.04	1.35	4.81
Total	549	2,433	2.07	1.19	78.34	15.18	4.68	1.37	3.80
F-statistic (ANOVA)	-	-	37.23 ***	14.1 ***	2.37 *	32.00 ***	47.19 ***	12.42 ***	30.37 ***

Note: TAD = transit-adjacent development; TOD = transit-oriented development.***: $p<.01$, **: $p<.5$, *: $p<.1$.

Table 4. Differences in Travel Behavior between Station Area Types after Matching

Area type pair	Travel outcomes	Observed difference	PSM ATE	ATE/observed difference ratio	Mean of control group	ATE/control ratio
TAD (control)–TOD	VMT	6.34***	5.49**	0.87	14.18	0.39
n = 564 (unmatched), *n* = 108 (matched)	Auto trips	4.02***	1.44**	0.36	4.06	0.35
	Transit trips	−0.64***	−0.74***	1.16	0.50	−1.48
	Walk trips	−2.86***	−1.89***	0.66	1.72	−1.10
Hybrid (control)–TOD	VMT	3.64***	0.48	0.13	9.53	0.05
n = 2,204 (unmatched), *n* = 350 (matched)	Auto trips	2.87***	0.84**	0.29	3.10	0.27
	Transit trips	0.11	−0.12	−1.09	1.39	−0.09
	Walk trips	−0.89***	−1.20***	1.35	3.95	−0.30
TAD (control)–Hybrid	VMT	2.70**	3.75*	1.39	23.69	0.16
n = 2,142 (unmatched), *n* = 364 (matched)	Auto trips	1.15***	0.00	0.00	6.70	0.00
	Transit trips	−0.75***	−0.35*	0.47	0.73	−0.48
	Walk trips	−1.97***	−0.75*	2.08	2.08	−0.36

Note: TAD = transit-adjacent development; TOD = transit-oriented development.***: $p<.01$, **: $p<.5$, *: $p<.1$ (*t*-test results).

With the six explanatory variables—household size, the number of workers, household income, distance to the nearest transit station, regional job accessibility, and the regions—household pairs in three area type pairs (TOD–TAD, TOD–Hybrid, and TAD–Hybrid) are matched. The PSM generates 54 household pairs (108 in total) in the TOD–TAD pair, 175 pairs in the TOD–Hybrid pair, and 182 pairs in the TAD–Hybrid pair.

After matching, whether the chosen residents in one type are systematically different from those in another type was evaluated. If they are different in terms of demographics, self-selection is still a concern. Unlike unmatched samples where all demographic variables are significantly different between pair groups, *t*-test results for matched samples show that residents in TOD and TAD do not differ by all covariates used in the PSM. Those variables are not statistically different in both TOD–Hybrid and TAD–Hybrid pairs as well (results are not shown).

Once the matching was complete, the ATE, the observed differences, and the ratio between them on VMT, auto trips, transit trips, and walk trips for each area pair were calculated. As an example for the TOD–TAD pair, the observed difference is the mean travel factors of all TOD households minus that of all TAD households in the original sample. The ATE is the difference in mean travel factors between the matched samples in TOD and TAD.

From the 3rd to 7th columns, Table 4 shows observed difference in mean in the original sample, ATE in matched sample, ratio of ATE over observed difference, mean value of control group after matching (the control group is TAD in the first and third pair and Hybrid in the second pair), and ratio of ATE over control mean, respectively. Regarding VMT in TOD–TAD pairs, after matching (i.e., controlling residential self-selection), TAD households tend to drive 5.49 miles per day more than TOD residents ("PSM ATE" column in Table 4). The significant effect on VMT of living in TOD accounts for approximately 87% of the observed influence ("ATE/observed difference ratio" column in Table 4); that is, 13% of the observed difference may result from residential self-selection.

The mean difference in automobile trips between matched TAD and TOD households is 1.44 and statistically significant. That is, if a randomly selected household moves from a TAD to a TOD, a decrease in the number of driving by 1.44 trips per day would be expected. The effect on auto trips of living in TOD accounts for approximately 36% of the observed influence. On average, the matched sample households in TAD drove 4.06 times per day. Thus, the effect of living in TOD itself represents a considerable 35% decrease (1.44 fewer trips) in daily auto trips.

In addition, the probability of walking or taking transit significantly decreases from TAD to TOD. After matching, an average household living in TOD takes 0.74 transit trips more (or 148% higher) than that in TAD. Likewise, the average household in TOD talks 1.89 walk trips more (or 110% higher) than that in TAD, and both differences are statistically significant.

When Hybrid and TOD areas are compared, the number of auto trips is significantly higher (0.84 more auto trips), and the number of walk trips is significantly lower (1.20 fewer walk trips) in Hybrid areas. The effect of living in TOD itself represents a 27% decrease in daily auto trips and a 30% increase in daily walk trips, compared with living in a Hybrid type. VMT and transit trips are not significantly different between Hybrid and TOD types.

In the case of the TAD–Hybrid pair, except auto trips which are not significant, VMT, walk trips, and transit trips are slightly significant. Compared with TAD areas, VMT is lower (3.75 more miles), the number of transit trips is higher (0.35 more transit trips), and the number of walk trips is higher (0.75 more walk trips) in Hybrid areas. The effect of living in the Hybrid type itself represents a 16% decrease in VMT, a 48% increase in daily transit trips, and a 36% increase in walk trips, compared with living in a TAD type.

From the results of travel outcomes between three pairs of station area type, the gradual and cumulative changes could be examined. When a household moves from a TAD type to a Hybrid type, or a local government improves a TAD to a Hybrid by increasing its density, land use mix, or walkability, the average household is expected to have slightly shorter auto trips and more walk trips and transit trips. Then if the household moves to a TOD, or the station area is developed to a TOD, the household is expected to have fewer auto trips and more walk trips. Cumulatively, from TAD to TOD, a household is estimated to have significantly shorter and fewer auto trips and more transit and walk trips.

Discussion and Conclusions

The clustering approach in this study classified existing station areas into TOD, TAD, and Hybrid types in terms of built environment factors—density, diversity, and street network design. As a result, 11% of the 549 stations in eight regions were labeled as TOD as being dense, diverse, and walkable. One-fifth were named as TAD as having opposite urban form of TOD. The other 70% of the stations were classified as Hybrid. Land use mix was a key factor to distinguish TAD from Hybrid, whereas density and street design played important roles to differentiate TOD and Hybrid.

Station area types vary among the literature according to classifying methods, factors, and regions. This study has an advantage as the area types are drawn from all stations in eight urban areas in the U.S., whereas the majority of studies are limited to one or few regions. Renne and Ewing (*14*) cover 54 regions across the U.S., but this study uses a more objective and systematic approach—hierarchical cluster analysis—and analyzes more travel outcome variables using disaggregate data.

Household characteristics and travel behaviors from household travel survey data were matched to each station area type, and this study found that residents living in different types are different from each other. Households in TAD tend to be more affluent, have more cars, live in a larger household, and be more auto-oriented than their counterparts in TOD. Regarding travel behavior, TAD households have much higher VMT and lower walk and transit trips than those in TODs and Hybrids. The average number of daily automobile trips shows the most dramatic differences, in that TAD households generate three times more auto trips than TOD households (6.06 vs. 2.04). The large difference in mode share between TOD and TAD (e.g., auto mode shares in TAD and TOD are 68% and 25%, respectively) is observed in other studies (*14*, *47*), some of which is much less dramatic—approximately 70% (TOD) vs. 85% (non-TOD) (*17*, *48*).

In this study, PSM enabled matching of samples to control for residential self-selection. Although the differences in travel outcomes become less dramatic after controlling self-selection, the matched sample still shows that TOD motivates its residents to walk more and take transit more while using personal vehicles less. The significant difference between TOD and TAD in both VMT and the number of auto trips means that TOD makes personal vehicle trips shorter (39% deduction) and fewer (35% deduction).

By considering the in-between hybrid type, this study may provide some practical implications. For example, when a local government and transit authority develop a TAD-type station area which is sprawled, single use, and not walkable, into a Hybrid type, mainly by adding different land uses, they could expect small increases in transit and walk trips. A Hybrid type of station area could be changed into a TOD type by adding density and decreasing block sizes, which would result in less driving and more walking by residents. The cumulative change from TAD to TOD could encourage its residents to drive less with shorter distances, walk more, and take transit more, which will have positive impacts on the city's environment, society, and economy.

This study has three main limitations. First, station area classification might generate different results if the input is changed; for example, if different regions or different built environment factors are included. The result also depends on the clustering method. The classification relying on cluster analysis is not able to provide specific guidance such as benchmark thresholds of density, land use mix, or street connectivity (for benchmark threshold examples, see Renne and colleagues [*14*, *49*]). Nevertheless, the clustering approach in this study might reflect reality better than hypothetical benchmarks do. Also, the result shows that different station areas such as TOD and TAD have meaningfully different effects on the travel outcomes of their residents.

Second, PSM works only when all confounding factors are included in the analysis. This study, however, only includes the factors reflecting self-selection indirectly, and does not have residents' attitude information. The risk of not controlling all confounding factors is that the effect of residential self-selection on travel behavior might be under- or over-estimated. To the authors' knowledge, there are no such attitude data covering multiple regions in the U.S., but the result of this study needs to be checked for external validity by additional TOD studies including residential preference data in specific regions.

Third, in theory, the observed covariates in the propensity score equation are measured before the treatment whereas the outcome is measured after the treatment (*44*). In the context of this study, the data collection point for household characteristics and location factors needs to be before the station area is developed, whereas the travel outcome data should be collected after the development. This requires longitudinal data. However, because the regional household travel surveys are conducted in different years in each region, it is not plausible to put all longitudinal data into one analysis. Although this study uses cross-sectional data to control the temporal differences across regions and stations, further research needs more advanced methods, such as natural experiments with longitudinal data.

Nevertheless, as first-of-its-kind research using both cluster analysis and PSM in the exploration of TOD/TAD outcomes, this study provides evidence that a TOD, and even a Hybrid type of station area, could encourage its residents to use more active modes of transportation. An effort to create a transit-oriented neighborhood does not have to be a "megaproject." Gradual changes of a station area into a denser, more diverse, and more walkable environment would compensate in the form of sustainable travel behavior, which ultimately provides more environmental, social, and economic benefits.

Acknowledgment

This research was funded by the National Institute for Transportation and Communities (NITC) under grant number 859, a program of the Transportation Research and Education Center at Portland State University and a U.S. Department of Transportation university transportation center.

Author Contributions

The authors confirm contribution to the paper as follows. Study conception and design: Keunhyun Park, Reid Ewing, Brenda Scheer; data collection: Keunhyun Park, Shabnam Khan; analysis and interpretation of results: Keunhyun Park, Reid Ewing; draft manuscript preparation: Keunhyun Park. All authors reviewed the results and approved the final version of the manuscript.

Note

1. The entropy index measures balance between three different land uses. The index ranges from 0, where all land is in a single use, to 1 where land is evenly divided among the three uses. Values are intermediate when buffers have more than one use but one use predominates. The entropy calculation is: entropy = – [residential share * ln(residential share) + commercial share * ln(commercial share) + public share * ln(public share)]/ln(3), where ln is the natural logarithm of the value in parentheses and the shares are measured in terms of total parcel land areas (*40*).

References

1. U.S. Bureau of Labor Statistics. *Consumer Expenditure Survey*. Inter-university Consortium for Political and Social Research, Ann Arbor, Mich., 2014.
2. Cervero, R. *Ridership Impacts of Transit-Focused Development in California*. Monograph 45. Institute of Urban and Regional Development, University of California, Berkeley, 1993.
3. Cervero, R. *TCRP Report 102: Transit-Oriented Development in the United States: Experiences, Challenges, and Prospects* Transportation Research Board of the National Academies, Washington, D.C., 2004.
4. Langlois, M., D. van Lierop, R. A. Wasfi, and A. M. El-Geneidy. Chasing Sustainability: Do New Transit-Oriented Development Residents Adopt More Sustainable Modes of Transportation? *Transportation Research Record: Journal of the Transportation Research Board*, 2015. 2531: 83–92.
5. Nasri, A., and L. Zhang. The Analysis of Transit-Oriented Development (TOD) in Washington, D.C. and Baltimore Metropolitan Areas. *Transport Policy*, Vol. 32, No. 1, 2014, pp. 172–179.
6. Olaru, D., and C. Curtis. Designing TOD Precincts: Accessibility and Travel Patterns. *Planning Practice and Research*, Vol. 15, No. 1, 2015, pp. 1567–7141.
7. Venigalla, M., and A. Faghri. A Quick-Response Discrete Transit-Share Model for Transit-Oriented Developments. *Journal of Public Transportation*, Vol. 18, No. 3, 2015, pp. 107–123.
8. Renne, J. L., T. Tolford, S. Hamidi, and R. Ewing. The Cost and Affordability Paradox of Transit-Oriented Development: A Comparison of Housing and Transportation Costs across Transit-Oriented Development, Hybrid and Transit-Adjacent Development Station Typologies. *Housing Policy Debate*, Vol. 26, No. 4–5, 2016, pp. 819–834.
9. Bernick, M., and R. Cervero. *Transit Villages in the 21st Century*. McGraw-Hill, New York, 1997.
10. Kamruzzaman, M., F. M. Shatu, J. Hine, and G. Turrell. Commuting Mode Choice in Transit Oriented Development: Disentangling the Effects of Competitive Neighbourhoods, Travel Attitudes, and Self-Selection. *Transport Policy*, No. 42, 2015, pp. 187–196.
11. Belzer, D., and G. Autler. *Transit Oriented Development: Moving from Rhetoric to Reality*. Brookings Institution Center on Urban and Metropolitan Policy, Washington, D.C., 2002.
12. Cervero, R., and M. Duncan. *Residential Self Selection and Rail Commuting: A Nested Logit Analysis*. Working Paper 604. University of California Transportation Center, Berkeley, 2002.

13. Dittmar, H., and G. Ohland. *The New Transit Town: Best Practices in Transit Oriented Development*. Island Press, Washington, D.C., 2004.
14. Renne, J. L., and R. Ewing. *Transit-Oriented Development: An Examination of America's Transit Precincts in 2000 & 2010*. UNOTI Publications Paper No. 17. University of New Orleans, 2013.
15. Laaly, S. *New Definition of Transit Oriented Development (TOD) Based on Context Sensitive Paradigm*. Doctoral dissertation. Morgan State University, 2014.
16. Pollack, S., A. Gartsman, M. Boston, A. Benedict, and J. Wood. Rating the Performance of Station Areas for Effective and Equitable Transit-Oriented Development. Presented at 93rd Annual Meeting of the Transportation Research Board, Washington, D.C., 2014.
17. Jeihani, M., L. Zhang, A. Ardeshiri, A. Amiri, A. Nasri, K. R. Zamir, and B. Baghaei. *Development of a Framework for Transit-Oriented Development (TOD)*. No. MD-13-SP209B4N. Maryland State Highway Administration Office of Policy and Research, Baltimore, 2013.
18. Canepa, B. Bursting the Bubble: Determining the Transit-Oriented Development's Walkable Limits. *Transportation Research Record: Journal of the Transportation Research Board*, 2007. 1992: 28–34.
19. Cervero, R., and K. Kockelman. Travel Demand and the 3Ds: Density, Diversity, and Design. *Transportation Research Part D: Transport and Environment*, Vol. 2, No. 3, 1997, pp. 199–219.
20. Cervero, R., and R. Gorham, R. Commuting in Transit versus Automobile Neighborhoods. *Journal of the American Planning Association*, Vol. 61, No. 2, 1995, pp. 210–225.
21. Vale, D. S. Transit-Oriented Development, Integration of Land Use and Transport, and Pedestrian Accessibility: Combining Node-Place Model with Pedestrian Shed Ratio to Evaluate and Classify Station Areas in Lisbon. *Journal of Transport Geography*, No. 45, 2015, pp. 70–80.
22. Ngo, V. Identifying Areas for Transit-Oriented Development in Vancouver. *Trail Six: An Undergraduate Journal of Geography*, Vol. 6, 2012, pp. 91–102.
23. Kamruzzaman, M., D. Baker, S. Washington, and G. Turrell. Advance Transit Oriented Development Typology: Case Study in Brisbane, Australia. *Journal of Transport Geography*, Vol. 34, 2014, pp. 54–70.
24. Brown, B. B., and C. M. Werner. The Residents' Benefits and Concerns Before and After a New Rail Stop: Do Residents Get What They Expect? *Environment and Behavior*, Vol. 43, 2011, pp. 789–806.
25. Werner, C. M., B. B. Brown, and J. Gallimore. Light Rail Use is More Likely on ''Walkable'' Blocks: Further Support for Using Micro-Level Environmental Audit Measures. *Journal of Environmental Psychology*, Vol. 30, No. 2, 2010, pp. 206–214.
26. Center for Transit-Oriented Development. *TOD 204 Planning for TOD at the Regional Scale: The Big Picture*. Report TOD 204. Reconnecting America, Washington, D.C., 2011.
27. Noland, R. B., K. Ozbay, S. Dipetrillo, and S. Iyer. *Measuring Benefits of Transit Oriented Development*. Mineta National Transit Research Consortium Report 12-18. Mineta National Transit Research Cosortium, Mineta Transportation Institute, College of Business, San Jose State University, 2014.
28. Center for Transit-Oriented Development. *Performance-Based Transit-Oriented Development Typology Guidebook*. Report TOD 204. Center for Transit-Oriented Development, Austin, 2010.
29. Cervero, R., and G. B. Arrington. Vehicle Trip Reduction Impacts of Transit-Oriented Housing. *Journal of Public Transportation*, Vol. 11, No. 3, 2008, pp. 1–17.
30. Cao, X. J., P. L. Mokhtarian, and S. L. Handy. Examining the Impacts of Residential Self-Selection on Travel Behaviour: A Focus on Empirical Findings. *Transport Reviews*, Vol. 29, No. 3, 2009, pp. 359–395.
31. Mokhtarian, P. L., and C. Xinyu. Examining the Impacts of Residential Self-Selection on Travel Behavior: A Focus on Methodologies. *Transportation Research Part B: Methodological*, Vol. 42, No. 3, 2008, pp. 204–228.
32. Handy, S. *Critical Assessment of the Literature of the Relationships among Transportation, Land Use and Physical Activity*. Resource paper for: Does the Built Environment Influence Physical Activity? Examining the Evidence—Special Report 282. National Academies Press, Washington, D.C., 2005.
33. Ewing, R., S. Hamidi, and J. B. Grace. Compact Development and VMT—Environmental Determinism, Self-Selection, or Some of Both? *Environment and Planning B: Planning and Design*, Vol. 43, No. 4, 2016, pp. 737–755.
34. Boer, R., Y. Zheng, A. Overton, G. K. Ridgeway, and D. A. Cohen. Neighborhood Design and Walking Trips in ten US Metropolitan Areas. *American Journal of Preventive Medicine*, Vol. 32, No. 4, 2007, pp. 298–304.
35. Cao, X. J. Exploring Causal Effects of Neighborhood Type on Walking Behavior Using Stratification on the Propensity Score. *Environment and Planning A*, Vol. 42, No. 2, 2010, pp. 487–504.
36. Cao, X. J., Z. Xu, and Y. Fan. Exploring the Connections among Residential Location, Self-Selection, and Driving: Propensity Score Matching with Multiple Treatments. *Transportation Research Part A: Policy and Practice*, Vol. 44, No. 10, 2010, pp. 797–805.
37. Cao, X. J., and Y. Fan. Exploring the Influences of Density on Travel Behavior Using Propensity Score Matching. *Environment and Planning B: Planning and Design*, Vol. 39, No. 3, 2012, pp. 459–470.
38. Cao, X. J., and J. Schoner. The Influence of Light Rail Transit on Transit Use: An Exploration of Station Area Residents along the Hiawatha Line in Minneapolis. *Transportation Research Part A: Policy and Practice*, Vol. 59, 2014, pp. 134–143.
39. Baghaei, B. *Impact of Transit Oriented Development on Mode Choice with Consideration for Multimodal Accessibility and Self-Selection*. Master of Science thesis, University of Maryland, College Park, 2014.
40. Ewing, R., G. Tian, J. P. Goates, M. Zhang, M. J. Greenwald, A. Joyce, J. Kircher, and W. Greene. Varying Influences of the Built Environment on Household Travel in 15 Diverse Regions of the United States. *Urban Studies*, Vol. 52, No. 13. 2015, pp. 2330–2348.
41. Atkinson-Palombo, C., and M. J. Kuby. The Geography of Advance Transit-Oriented Development in Metropolitan Phoenix, Arizona, 2000–2007. *Journal of Transport Geography*, Vol. 19, No. 2, 2011, pp. 189–199.

42. Charrad, M., N. Ghazzali, V. Boiteau, A. Niknafs, and M. M. Charrad. NbClust: An R Package for Determining the Relevant Number of Clusters in a Data Set. *Journal of Statistical Software*, Vol. 61, No. 6, 2014, pp. 1–36.
43. Oakes, J. M., and P. J. Johnson. Propensity Score Matching for Social Epidemiology. *Methods in Social Epidemiology*, No. 1, 2006, pp. 370–393.
44.Rosenbaum, P. R., and D. B. Rubin. Reducing Bias in Observational Studies Using Subclassification on the Propensity Score. *Journal of the American Statistical Association*, Vol. 79, No. 387, 1984, pp. 516–524.
45Austin, P. C. Optimal Caliper Widths for Propensity-Score Matching When Estimating Differences in Means and Differences in Proportions in Observational Studies. *Pharmaceutical Statistics*, Vol. 10, No. 2, 2011, pp. 150–161.
46.Lechner, M. Program Heterogeneity and Propensity Score Matching: An Application to the Evaluation of Active Labor Market Policies. *Review of Economics and Statistics*, Vol. 84, No. 2, 2002, pp. 205–220.
47. Renne, J. *Transit Oriented Development: Measuring Benefits, Analyzing Trends, and Evaluating Policy*. PhD Dissertation. Rutgers University, New Brunswick, NJ, 2005.
48. Kamruzzaman, M., D. Baker, S. Washington, and G. Turrell. Residential Dissonance and Mode Choice. *Journal of Transport Geography*, Vol. 33, 2013, pp. 12–28.
49. Renne, J. L., T. Tolford, S. Hamidi, and R. Ewing. The Cost and Affordability Paradox of Transit-Oriented Development: A Comparison of Housing and Transportation Costs Across Transit-Oriented Development, Hybrid and Transit-Adjacent Development Station Typologies. *Housing Policy Debate*, Vol. 26, No. 4–5, 2016, pp. 819–834.

The Standing Committee on Transportation and Land Development (ADD30) peer-reviewed this paper (18-00768).

Article

TRR
JOURNAL OF THE TRANSPORTATION RESEARCH BOARD

Transportation Research Record
2018, Vol. 2672(6) 40–51

DOI: 10.1177/0361198118788423
journals.sagepub.com/home/trr
SAGE

Impact of Built Environment on First- and Last-Mile Travel Mode Choice

Baichuan Mo[1,2], Yu Shen[3,4], and Jinhua Zhao[5]

Abstract

The paper studies the impacts of built environment (BE) on the first- and last-mile travel modal choice. We select Singapore as a case study. The data used for this work is extracted from the first- and last-mile trips to mass rapid transit (MRT) stations in the Household Interview Travel Survey of Singapore in 2012 with nearly 24,000 samples. The BE indicators are quantified based on four "D" variables: Density, Diversity, Design, and Distance to transit. We also take into account sociodemographic and trip-specific variables. Mixed logit (ML) modeling frameworks are adopted to estimate the impact of BE and the heterogeneity of taste across the sample. Based on the availability of light rail transit (LRT) in different areas, two modeling structures are implemented with binary ML models for non-LRT areas where "walk" and "bus" are the available travel modes, and multinomial ML models for areas where LRT is an additional alternative. The modeling results shed light on the following findings: BE—especially distance to MRT station, transportation infrastructures, land-use mix, and socioeconomic activities—significantly influences the first- and last-mile travel behaviors. Those who live or work close to MRT stations and in an area with high socioeconomic activities and land-use mix may have stronger preferences to walk for the first- and last-mile trips. The impact of physical BE (i.e., distance, infrastructures) is relatively homogeneous among the sample, while the impact of socioeconomic BE factors (i.e., floor space density, entropy) tends to vary across the sample.

Built environment (BE) is the man-made space in which people live, work, and recreate on a day-to-day basis (*1*). It encompasses urban design, land use, transportation system and patterns of human activity within the physical environment (*2*). BE can be quantified in several ways. One of the most widely used definitions is the well-known "D" variables of Ewing and Cervero (*3*). Past studies have revealed the impacts of urban form (*4–6*) and BE (*3, 7, 8*) on travel behavior, from which the findings provide profound reference for urban planning policy. Focusing on access to and egress from transit facilities—so-called first- and last-mile trips—studies on the influence of BE on first- and last-mile travel behaviors are, however, few. Cervero et al. (*9*) found that people in places with denser BE usually walk to transit stations. Similar conclusions are also drawn by Daniels and Mulley (*10*): walking distance to transit stops is mostly related to the mode of transit being accessed. Looking into the influence of street design and walkability on travel mode choice to transit stations, Park (*11*) yielded that better walkability increases the probability of transit users choosing to walk instead of driving to the stations. The BE factors that these studies take into account are, however, not sufficiently complete, and more comprehensive analysis on the relationship between BE and the first- and last-mile travel behavior can hardly be found. Tilahun et al. (*12*) conducted a wider range analysis of the last-mile issues in commuting trips incorporating the impact of BE. Nevertheless, by restricting the study to commuting trips, the findings may probably cause some bias in estimating travel behavior. Traffic conditions and demographic characteristics vary by country. The results of previous studies in America or Europe may not be suitable for Asian countries like Singapore. To fill the research gap, this study presents a comprehensive analysis on the impact of BE on first- and last-mile travel mode choice in Singapore, with the four "D" characteristics—Density, Diversity, Design and

[1]Department of Civil and Environmental Engineering, Massachusetts Institute of Technology, Cambridge, MA
[2]Department of Civil Engineering, Tsinghua University, Beijing, China
[3]Key Laboratory of Road and Traffic Engineering of the Ministry of Education, Tongji University, Shanghai, China
[4]Future Urban Mobility IRG, Singapore-MIT Alliance for Research and Technology, Singapore
[5]Department of Urban Studies and Planning, Massachusetts Institute of Technology, Cambridge, MA, USA

Corresponding Author:
Address correspondence to Yu Shen: yshen@tongji.edu.cn

Distance to transit (3)—to capture different perspectives of BE. In particular, the heterogeneity of the impact of BE, which is seldom considered in the literature, is also studied in this paper.

The first- and last-mile problem is a challenge to promoting greater patronage of public transit. The distances between transit stations and the origins/destinations of passengers may sometimes be greater than they are willing to walk. One has to choose a feeder travel service to reach the transit station or even use an alternative direct travel mode like personal vehicles, resulting in the systematic decrease of accessibility in urban areas. Some solutions have been proposed, such as altering the location of transit stations to mixed-used activity centers; siting houses/workplaces near rail stations for improved proximity; and constructing pedestrian footways, shaded corridors, and bike lanes to improve walkability and connectivity (9, 12–16). Most of these solutions tend to redesign or adjust the BE to improve the first- and last-mile experience.

In this study, we investigate the impact of the BE on first- and last-mile modal choice. A mixed logit (ML) model framework is used to capture the heterogeneity of the impact of BE, which contributes to current literature in addition to previous studies (9–12). Singapore is selected as the case study area. Residents in Singapore rely heavily on public transport for daily travel. According to the Household Interview Travel Survey (HITS) in 2012, during morning peak hours, 70% of commuters go to work by public transit, including mass rapid transit (MRT) and bus. Thus, the first- and last-mile problem cannot be neglected in Singapore. In addition, the modal share of the first- and last-mile trips varies across the MRT stations (17), which may reflect the influence of BE in various locations. Such circumstances raise the importance in understanding the roles that BE plays on daily travel behaviors, especially in the context of Singapore.

The rest of this paper is organized as follows. The next section presents the processing and descriptive analysis of data. The methodology and model results are described in the third section. The final section discusses the findings and concludes the study.

Data Processing and Descriptive Analysis

In the study, three categories of variables that may influence the first- and last-mile travel behaviors are collected. They are classified as sociodemographic variables (e.g., income, gender), trip-specific variables (e.g., travel time, travel cost), and BE variables. The data processing methods and descriptive analysis are illustrated in the following.

Household Interview Travel Survey

HITS is a paper-based household survey conducted every five years with a special focus on travel behavior in Singapore. The survey collects data on travel characteristics as well as individual sociodemographic information. The survey targets a sample size of at least 10,000 households, about 1% of the total number of households in Singapore (18). The sampled households are randomly selected by computer programs to ensure the representativeness of population. The data are collected through face-to-face interviews. The survey method follows the standard trip diary-based approach. In HITS, a trip is defined as a one-way journey completed for a specific purpose. On average 2.4 trips are collected for each respondent. The trip-specific characteristics of each travel segment (e.g., walking time to a bus stop, travel mode from home to the MRT station) are also recorded, which allows the identification of first- and last-mile trips.

In this study, a first/last mile trip is defined as the trip between an MRT station and the origin or destination of the journey. All travel records with MRT segments are first extracted from the HITS data. Then, the trip segments before and after the MRT trip are separated from the extracted records as the first- and last-mile samples. The sociodemographic information and trip-specific characteristics are collected as well. Samples with travel distance greater than 3 km are excluded, due to the fact that they are usually beyond the maximum service distance of an MRT station. These observations are not taken into account in the first/last trips in this study. The exclusion of these data has little impact on the modal share of samples. In total 23,941 trips are extracted from the HITS database. The characteristics of travel segments are only recorded if a mode is actually used in the trip. We use Google Maps API to calculate the travel time and cost of the same trip by alternative modes, based on the departure time recorded in HITS.

Built Environment

The BE data are derived from the Singapore Land Authority digitized cadastral dataset and the synthetic population data described by Zhu and Ferreira (19). The former contains detailed BE information, such as land use, postal codes, and survey district numbers and boundaries. The latter are computed based on the iterative proportional fitting with two-stage approach (19), which contains the quantity and location of employment, residents, and building floor space.

We divide Singapore into 1,169 zones on the basis of traffic analysis zones (MTZs) as shown in Figure 1. The average size of each MTZ is about 0.93 km^2. According to Ewing and Cervero (7), the BE impact is often studied in the neighborhood or activity center level in the literature. Thus, these divisions are reasonable for the BE variables calculation.

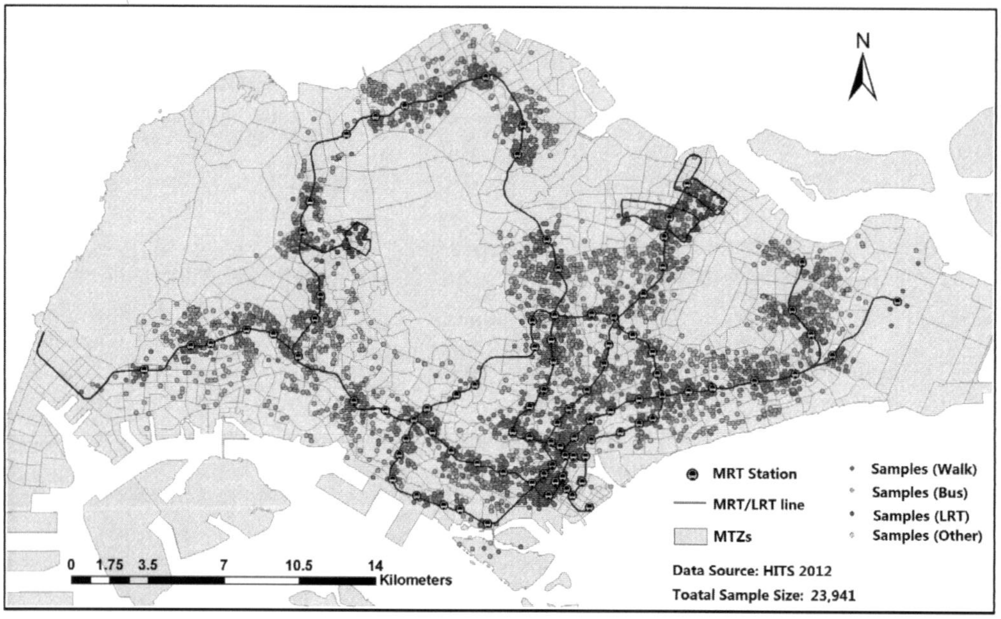

Figure 1. Distribution of samples in MTZs.

In this study, four "D" indicators are used to represent the BE: density, design, diversity, and distance to transit (*3*). Table 1 elaborates the measurement of each "D" variable. The variables are calculated in ArcGIS. Population density is defined as the total number of residents living in each MTZ divided by the corresponding residential area. The density of employment is estimated based on the methods proposed by Munshi (*20*), described as the total number of jobs in an MTZ divided by the economic activity area (e.g., commercial, industrial). The floor space density is calculated as the total building floor space in a specific MTZ divided by the area of this zone.

In terms of the diversity variables, we calculated the dissimilarity index and entropy based on Cervero and Kockelman (*21*). Six land-use categories are classified: residential, commercial, institutional, industrial, recreational, and other (e.g., waterbody). We first latticed the island of Singapore into 100 x 100 meter grid cells. The cells are set as the basic unit to calculate the diversity indices. Each grid cell is labeled with its most prevalent land use in terms of gross floor area for dissimilarity index calculation. A value of dissimilarity is then computed for each cell as the number of dissimilar land uses in labels of the eight "queen" neighborhood cells. The dissimilarity index of a MTZ is computed as the mean dissimilarity values of all internal cells. Greater dissimilarity index indicates higher level of land use mixture, typically considered as characteristics of smart growth *22*). To calculate the entropy, a buffer of 800-meter radius around the center of each grid cell was considered as the neighborhood area (*20*, *21*). The entropy of each grid cell is then estimated based on the land-use categories within the buffer. Similarly, the entropy of an MTZ is computed as the mean of the entropy value of all internal grid cells. The entropy index ranges from 0 to 1, where 1 signifies the perfect balance of land use with maximum heterogeneity and 0 indicates that there is only one land use in the neighborhood area (*21*).

The category of design is represented by the density of road length and road intersections, the kernel density of bus stops and MRT/LRT stations, and the ease of access index (EAI) to bus stops, MRT/LRT stations, and buildings. The density of road intersections represents the complexity of the road network and size of blocks. Expressways and walking paths are excluded from road length density calculation since they are seldom used by vehicles for first- and last-mile trips. The road length density is expected to have a positive effect on the choice of motorized travel modes but a negative effect on walking since high road length density means greater connectivity to motor vehicles (*24*). A kernel radius—2 km for bus and LRT stations and 3 km for MRT stations—and a distance decay function are assigned to each bus stop or MRT station for kernel density calculation (*22*). The value of the decay function reaches the peak at the source point and decreases smoothly as the distance increases within the kernel radius. The kernel value for each grid cell was computed as the sum of all overlapped function values. For each MTZ, the kernel density is calculated as

Table 1. Definition of BE Variables

Variable		Definition	Operationalization
Density	Population density	Population per residential square meter.	$D_p = P_r/A_r$, where D_p= population density, P_r = population in the residential area, A_r = area of residential land use in a MTZ.
	Employment density	Employment per residential square meter.	$D_e = E_{ea}/A_{ea}$, where D_e= employment density, E_{ea} = total employment available in the economic activity area, A_r = area of economic activity land use in a MTZ.
	Floor space density	Building floor space per square meter.	$D_{fs} = FS/A$, where D_{fs} = floor space density, FS = total floor space in a MTZ, A = area of the MTZ.
Diversity	Dissimilarity index	Proportion of dissimilar land use among grid cells within a MTZ (21).	Dissimilarity index $= I_d = \sum_{k=1}^{K}\left[\sum_{l=1}^{8}(X_l/8)\right]/K$, where K = the number of grid cells in a MTZ, $X_l = 1$ if the land use category of neighboring grid cell l differs from the grid cell k.
	Entropy	The mean entropy for land use categories among grid cells within an 800-meter radius of each grid cell within a MTZ (21).	Entropy $= E = \left\{\sum_{k=1}^{K}\left[\sum_{j} P_{jk}\ln(P_{jk})\right]/\ln(J)\right\}/K$, where K = the number of grid cells in a MTZ, J = the number of land-use classes. P_{jk} is the proportion of the area under the j th land-use type within the 800-m radius surrounding grid cell k.
Design	Road length density	Road length per square meter.	Road length density $= D_r = L_{road}/A$, where L_{road} = road length, A = area of a MTZ.
	Road intersections density	The number of road intersections per square meter.	Road intersections density $= D_i = N_i/A$, where N_i = number of road intersections. A = area of a MTZ.
	Bus/MRT/LRT stations: kernel density	Mean kernel density for Bus/MRT/LRT stations (22).	Kernel density $= KD = \left[\sum_{k=1}^{K}\left(\sum_{m} e^{-d_{km}}\right)\right]/K$, where d_{km} = the distance (km) from grid cell k to Bus/MRT/LRT stations, K = the number of grid cells in a MTZ.
	EAI to MRT/LRT station	Individual's ease of access to surrounding MRT/LRT stations based on the gravity model (23).	EAI to MRT/LRT station $= \sum_{m} A_{sm}e^{-d_{im}}$ where d_{im} = the distance (km) from individual i to MRT/LRT stations m within 2 km, A_{sm} = the footprint ($1000\times m^2$) of the MRT/LRT station m.
	EAI to bus stop	Individual's ease of access to surrounding bus stops based on the gravity model. (23).	EAI to bus stop $= \sum_{m} A_{bm}e^{-d_{im}}$ where d_{im} = the distance (km) from individual$i$ to MRT/LRT stations m within 2 km, A_{bm} = the bus bay length (km) of the bus stop m.
	EAI to buildings	Individual's ease of access to surrounding buildings based on the gravity model. (23).	EAI to buildings $= \sum_{m} FS_{bm}e^{-d_{im}}$ where d_{im} = the distance (km) from individual i to building m within 2 km, FS_{bm} = the floor space ($1000\times m^2$) of the building m.
	Walking-based EAI to MRT station	Individual's ease of access to surrounding MRT stations by walking based on the gravity model (23).	Walking-based EAI to MRT station $= \sum_{m} A_{sm}e^{-T_{im}}$ where T_{im} = the walking time ($1000\times$sec.) of individual i to MRT station m within 2 km, A_{sm} = the footprint ($1000\times m^2$) of the MRT station m.
Distance to transit stop	Distance to MRT station	Euclidean distance between MRT station and the origin or destination.	Measured as Euclidean distance between MRT station and the origin or destination.

Note: EAI = ease of use index.

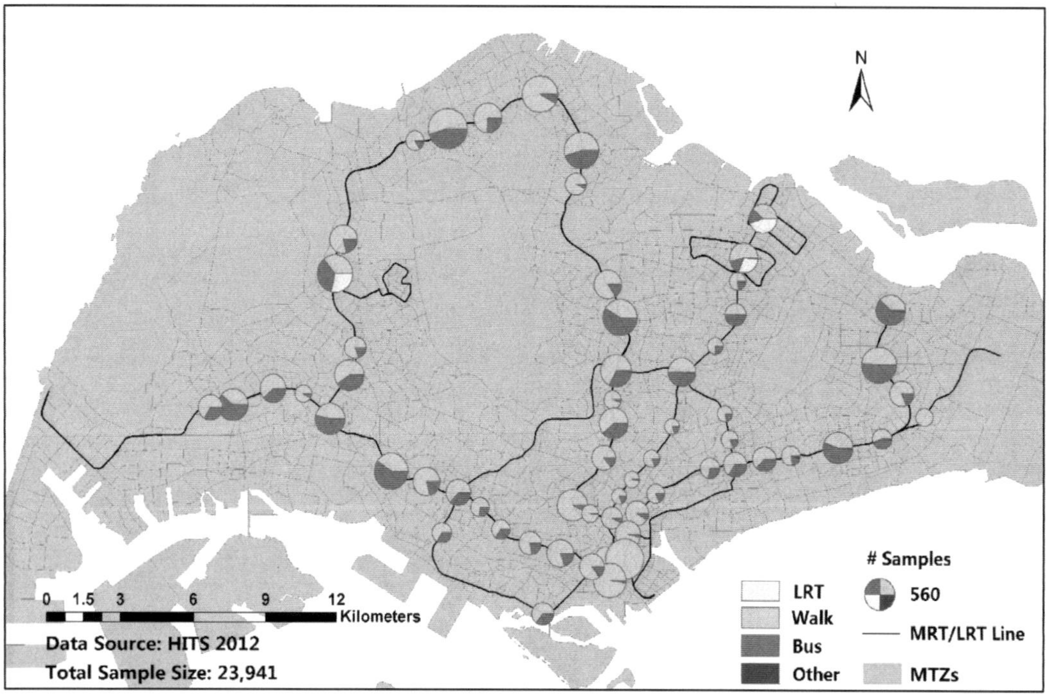

Figure 2. Modal share of first- and last-mile trips.

the average kernel value for all grid cells in the zone. As opposed to point density, kernel density has the advantage of counting the effect of BE not only on the zone containing the corresponding bus stops and MRT/LRT stations, but also on the neighborhood. Similar to the definition of Zhao (*24*), we define the EAI to represent the intensification of public transport services around an individual. It is the accumulation of the decay function value multiplied by footprint (i.e., the horizontal projection of the building) of the transit stops/stations within a 2 km distance threshold. The EAI to buildings is also defined by the same method, reflecting the potential of an individual to access surrounding socioeconomic activities, since the EAI to MRT stations is expected to play an important role on first- and last-mile trips. Two decay functions are used to calculate the EAI to MRT stations (decayed by distance) and the walking-based EAI to MRT stations (decayed by walking time), respectively. The distance to transit stop is calculated as the direct distance between MRT station and the origin/destination. It is expected to be the most significant factor for people choosing between walking and motorized travel modes.

Descriptive Analysis

The origins and destinations of the sample of 23,491 trips are illustrated in Figure 1, which covers nearly all surrounding regions of the MRT lines, showing the spatial representativeness of the samples. Most of the green dots (choosing walk) are located near the MRT stations while the majority of yellow dots (choosing bus) are more distant from MRT stations, which indicates that the distance to MRT stations plays an important role in the first- and last-mile travel mode choice.

In Singapore, "walk" and "bus" are the two major travel modes for first- and last-mile trips: 69.06% of the respondents choose to walk to or from MRT stations, while the share of bus mode is 27.75%. The LRT mode accounts for 2.21% and the others just 0.98%. The "other" modes (e.g., car driving and sharing) are not considered separately as the share is too small to estimate the variables' impacts. Moreover, according to numerical testing, discarding these modes has almost no influence on the modeling results. The small share of car usage may result from the fact that less than 10% of people in Singapore own private cars (*25*). Those who choose to drive are more likely to drive to the destination directly instead of driving to an MRT station for the first/last mile only. Figure 2 presents the first- and last-mile modal share around different MRT stations. The modal share varies across MRT stations. The finding motivates us to explore the implied impact of various types of BE on the first- and last-mile travel behaviors. The LRT is only available around three specific MRT stations (Choa Chu Kang, Sengkang, and Punggol). Therefore, two separate sample sets were created to model and to explore the

Table 2. Summary of Sociodemographic Variables

Variable name	Mean	SD
Live in public flat (Yes=1)	0.87	0.33
Singapore citizen (Yes=1)	0.83	0.38
Number of people in household	4.12	1.44
Have children under six (Yes=1)	0.15	0.36
Age below 25 (Yes=1)	0.30	0.46
Age between 25 and 45 (Yes=1)	0.44	0.50
Gender (Male=1)	0.48	0.50
Hold a driver's license (Yes=1)	0.29	0.45
Full-time employment (Yes=1)	0.89	0.31
Income ($SG)	1842.00	1935.63
Commute trip (Yes=1)	0.43	0.50

travel behaviors in areas with and without LRT infrastructures, respectively.

Tables 2–4 list a summary of the variables considered in this study. In terms of sociodemographic variables, the distributions of gender (48% male) and age (30% below age 25; 44% age 25–45 years) in the sample are reasonable. The BE variables of origin, destination, and non-MRT station area of each first- and last-mile trip are calculated. For a first-mile trip, the origin refers to traveler's home or workplace and the destination means the corresponding MRT station. For a last-mile trip, the origin becomes the MRT station while the traveler's home or workplace is the destination. In addition, the "non-MRT station area" represents the origin of the first mile and/or the destination of the last mile. The characteristics of socioeconomic activity of an individual can be reflected in these areas. Since the first- and last-mile trips are mostly made within a short distance, the three categories of BE variables are highly correlated with each other. Therefore, only one of them is selected in the final model on the basis of goodness-of-fit and explanatory reasonability. The mean entropy of samples is relatively high (0.60), indicating higher land-use balance of areas near MRT stations. A wide variation in kernel density of transit stops can be observed, indicating the spatial differences in the construction of public transport facilities in Singapore. In terms of the trip-specific variables, we can find that one tends to choose motorized travel modes in relatively long distance trips but to choose to walk if the distance is shorter. The mean travel time of first- and last-mile trips is thus well controlled in a reasonable range (about 7–10 min).

Table 3. Summary of BE Variables

Variable name	Mean	SD
Non-MRT station area*		
Population density (people/m^2)	0.059	0.028
Employment density (jobs/m^2)	0.14	0.20
Floor space density (m^2/m^2)	2.51	3.73
EAI to bus stop	0.97	0.24
EAI to MRT station	17.49	21.62
EAI to building	6296.29	2899.25
EAI to LRT station	0.31	1.08
Walking-based EAI to MRT station	32.73	47.77
Bus stops kernel density	1.20	0..31
MRT stations kernel density	20.54	23.57
LRT stations kernel density	0.43	1.24
Entropy	0.60	0.15
Dissimilarity index	0.25	0.15
Road density (0.01m/m^2)	1.55	0.94
Road intersections density (no./km^2)	17.51	18.37
Distance to MRT station (m)	713.35	623.66
Origin		
Population density (people/m^2)	0.065	0.34
Employment density (jobs/m^2)	0.50	0.20
Floor space density (m^2/m^2)	2.46	3.60
EAI to bus stop	1.00	0.25
EAI to LRT station	0.29	0.99
Bus stops kernel density	1.23	0.30
MRT stations kernel density	21.86	23.79
LRT stations kernel density	0.45	1.01
Entropy	0.61	0.15
Dissimilarity index	0.23	0.15
Road density (0.01m/m^2)	1.51	0.87
Road intersections density (no./km^2)	19.39	18.87
Destination		
Population density (people/m^2)	0.064	0.33
Employment density (jobs/m^2)	0.50	0.20
Floor space density (m^2/m^2)	2.47	3.62
EAI to bus stop	0.96	0.26
EAI to LRT station	0.34	1.17
Bus stops kernel density	1.20	0.32
MRT stations kernel density	20.54	23.57
LRT stations kernel density	0.43	1.24
Entropy	0.60	0.15
Dissimilarity index	0.25	0.15
Road density (0.01m/m^2)	1.51	0.87
Road intersections density (no./km^2)	19.30	18.60

Model Results and Analysis

Methodologies

The choice behavior, as stated by Train (*26*), is based on the rational evaluation of all available alternatives and the maximization of utility. In this study, the alternatives consist of walk, bus, and LRT. In addition to the BE variables, individual and household level factors are also assumed to affect the modal choice. Variables including gender, age, household size, income, travel purpose, etc. are all tested to capture the influence of these variables on modal choice. The trip-specific variables (e.g., travel time and travel cost) are also taken into account in the models. In summary, we assume that the first- and last-mile modal choice is influenced by three different categories of factors: sociodemographic characteristics of respondents, BE at the origin, destination and non-MRT station areas, and trip-specific variables.

Table 4. Summary of Market Share and Trip-Specific Variables

Model	Mode	Modal share (%)	Travel distance (m)		Travel time (min)	
			Mean	SD	Mean	SD
	Walk	72.30	655.39	456.41	7.47	3.93
Binary ML	Bus	26.74	2381.42	1229.09	10.34	5.14
	Other	0.96	na	na	na	na
Multinomial ML	Walk	52.74	763.16	432.05	6.97	3.92
	Bus	29.96	2045.41	985.53	8.87	4.19
	LRT	15.81	2642.75	1378.14	8.30	5.61
	Other	1.49	na	na	na	na

Note: na = not applicable; SD = standard deviation.

*Some BE variables (e.g., EAI to MRT station) of non-MRT station areas are not included in those of origin and destination since they are considered meaningless for first- and last-mile trips. For instance, the origin of a trip can be an MRT station. Thus, the EAI of an MRT station to another MRT station (i.e., EAI of origin to MRT station) is not meaningful for first- or last-mile trips.

The influence of BE on mode choice may vary across gender, household composition, and socioeconomic groups (*27*, *28*). A household with an inclination toward a certain type of travel may self-select a residential location to use the preferred mode to travel (*28*). This phenomenon is commonly referred to as the residential self-selection problem. In this study, the impact of self-selection bias is assumed to be mitigated due to the Build-to-Order (BTO) policy in Singapore (*29*). People who want to buy a public flat developed by the Housing and Development Board need to apply and wait to be allocated. This policy results in individuals having low flexibility in their choice of residence, as well as mixed sociodemographic characteristics in a residential building, mitigating the self-selection bias.

The ML model is a highly flexible model that allows for estimating the random taste variation across individuals (*26*). The heterogeneity of the impact of BE has seldom been estimated in past works. In this study, to estimate the mean impact and taste variation of BE, the ML modeling framework is adopted. Since the availability of LRT is constrained in three MRT station areas, individuals in these areas with EAI to LRT are selected to perform the multinomial ML model with three alternatives (i.e., walk, bus, and LRT). Other samples are selected to perform a binary ML model (with walk and bus as alternatives). According to discrete choice theory, individuals are rational and choose the travel mode that provides the highest utility. The utility function V_{ni} for person n choosing travel mode i is specified as follows (*26*, *30*):

$$V_{ni} = \alpha_i + \gamma_i' X_n + \theta_{ni}' B_n + \mu_{ni}' T_{ni} \tag{1}$$

where:

X_n is the vector of sociodemographic characteristics of individual n.

B_n is the vector of BE variables of individual n.

T_{ni} is the vector of trip-specific attributes of mode i for individual n.

α_i is the alternative specific constant (ASC), capturing the inherent preference for mode i.

γ_i', θ_{ni}' and μ_{ni}' are the corresponding coefficients to be estimated.

According to the classic theory of ML modeling estimation (*26*, *30*, *31*), the probability of individual n choosing travel mode i can be expressed as:

$$P_{ni} = \int \frac{\exp(V_{ni})}{\sum_{k=1}^{K} \exp(V_{nk})} f(\beta) d\beta \tag{2}$$

where:

β is the vector of coefficients which combines θ_{ni}' and μ_{ni}'.

P_{ni} is the probability for individual n choosing mode i, simulated by taking draws of β and from the density $f(\beta)$, which is assumed to be normal distribution in this study.

K is the number of available alternatives for the individual.

Since we focus on the impact of BE and aim to improve the estimation efficiency of the proposed model, the taste variation of sociodemographic variables and ASCs are not measured in this study.

For the purposes of analysis, aggregate elasticity is often used to summarize the impact of an incremental change in a variable on the expected share of a group of decision makers (*30*). Derived from Train (*26*) and Ben-Akiva and Lerman (*30*), the percentage change in the expected probability for alternative i given a percentage change in the mth attribute of alternative j in population level is:

$$E_{x_j^m}^{\bar{P}_i} = \frac{1}{N \cdot \bar{P}_i} \sum_{n=1}^{N} - x_{nj}^m \int \beta^m \frac{\exp(V_{ni})}{\sum_{k=1}^{K} \exp(V_{nk})} \frac{\exp(V_{nj})}{\sum_{k=1}^{K} \exp(V_{nk})} f(\beta) d\beta, \tag{3}$$

Table 5. Modeling Estimation Results of the Binary ML model

Variable		(a) Model with BE variables		(b) Model without BE variables	
		Coefficient	*t*-test	Coefficient	*t*-test
Walk					
Constant α	-	0	fixed	0	fixed
Travel time	Mean	–0.567	–20.06***	–0.580	–41.09***
	†SD	–0.115	0.11	0.383	0.23
Bus					
Constant α	-	–9.510	–17.25***	–6.38	–46.13***
Travel time	Mean	–0.946	–16.09***	–0.253	–21.73***
	SD	0.250	12.42***	0.058	4.97***
Commute trip (Yes=1)	-	0.235	1.81*	0.243	3.81***
Distance to MRT station	†Mean	1.160	15.95***	-	-
	‡SD	0.102	0.04	-	-
EAI to bus stop (origin)	Mean	2.650	6.98***	-	-
	SD	0.037	0.28	-	-
Floor space density (non-MRT station area)	Mean	–0.329	–4.53***	-	-
	SD	0.146	3.23***	-	-
Walking-based EAI to MRT station	Mean	–0.039	–6.43***	-	-
	SD	0.027	6.81***	-	-
Road density (non-MRT station area)	Mean	0.144	1.75*	-	-
	SD	0.362	0.06	-	-
Statistics					
Observations		20181		20181	
Rho squared		0.832		0.736	
Adjusted Rho squared		0.831		0.735	

†Data scaled by 100.
‡Data scaled by 10,000.
***Significant at 99% level.
**Significant at 95% level.
*Significant at 90% level.

where:

N is the number of samples.

x_{nj}^m is the mth attribute of alternative j for person n.

β^m is the mth attribute element of β.

$\bar{P}_i$ is expected possibility of the group choosing alternative i, satisfying that:

$$\bar{P}_i = \frac{\sum_{n=1}^{N} P_{ni}}{N}. \quad (4)$$

$E_{x_j^m}^{\bar{P}_i}$ is the aggregate elasticity of alternative i given the same increment of x_{nj}^m for each individual so that:

$$\frac{\partial x_{nj}^m}{x_{nj}^m} = \frac{\partial x_{n'j}^m}{x_{n'j}^m} = \frac{\partial x_j^m}{x_j^m}, \text{for all } n, n' = 1, 2, \ldots, N. \quad (5)$$

where

$$x_j^m = \frac{\sum_{n=1}^{N} x_{nj}^m}{N}. \quad (6)$$

The models were estimated in PythonBiogeme with 1,000 random draws (*32*). The base category of the model is "walk" and all other modes are studied in comparison with it. Some variables described in the previous section were dropped from the final model due to the fact that they were found to be insignificant in explaining mode choice or were highly correlated with variables that were eventually included in the model. If two variables are highly correlated, the selection of variables is based on the goodness-of-fit and the sensitivity of policy, that is, we opted to keep variables that are explanatory and constructive for policy.

Binary ML model

After filtering out the samples with the "other" travel modes, in total 20,181 samples are used for the binary ML model estimation. Apart from the normal model, for comparative purposes, a control model without BE variables is tested. The estimation results of model with and without BE variables are shown in Table 5.

Both models in the table reveal high goodness-of-fit values, indicating the models are able to predict the travel behavior well. As shown in the table, by incorporating the BE variables, the goodness-of-fit value, the adjusted ρ^2 substantially improves from 0.733 to 0.832, which indicates the importance of BE in addition to the trip-specific and sociodemographic variables.

As in Table 5 column (a), the signs of all coefficients are consistent with the assumed effects. The ASC of "bus" (–9.510) is less than "walk" (fixed to 0), which suggests a negative preference for bus travel. Compared with trip-specific and BE variables, sociodemographic variables have little effect on the first- and last-mile modal choice, which implies that the impact of self-selection could be minor. It is found that individuals traveling for the purpose of commuting (e.g., work and education) are more likely to choose bus. This suggests that the previous research (*12*) which studied the impact of BE on commuting trips may cause some bias. The model also suggests that travel time is an important consideration in the choice of modes. Referring to the estimates in Model (a), the absolute value of the coefficient of "bus" travel time is greater than that of "walk", indicating that individuals are more sensitive to time spent traveling by bus than time spent walking. The inconsistent values in Model (b) are possibly due to the fact that the parameter of bus travel time in Model (b) contains the hidden positive effect of distance to MRT station. Thus, after adding the influence of distance in Model (a), the impact of bus travel time is adjusted. The standard deviation of bus travel time is statistically different from zero, which shows varying attitudes toward bus travel time among the sample.

In terms of the impact of BE, distance to MRT station plays an essential role in modal choice. The positive sign demonstrates higher probability of choosing "bus" with longer distance. By calculating the aggregate elasticity based on Equation 3, if the distance to MRT station decreases by 10%, the share of "walk" on average tends to increase by 2.04% while the share of "bus" tends to decrease by 5.35%. In addition, the density of building floor space in non-MRT areas is found to have a negative effect on bus mode choice. This implies that higher density of socioeconomic activities encourages people to walk. Meanwhile, modal choice is less sensitive to road length density. If the road density increases by 10%, then the share of "walk" decreases by 0.07% while "bus" increases by 0.17% on average. The only BE parameters found to vary across individuals were walking-based EAI to MRT station and floor space density. The former indicates the variation of tastes on walking in the sample. Since floor space density is associated with socioeconomic prosperity, the latter suggests that the impact of the attractiveness of socioeconomic activities, such as business and recreational activities, varies among the sample. These two variables are both related to human activities; we may find that the impact of socioeconomic-related BE factors tends to vary across the sample. However, the insignificant standard deviation of other parameters (e.g., distance to MRT station, road length density) indicates more homogeneous impacts of physical BE on the sample.

Multinomial ML Model

To model the impact of BE in the areas with LRT, the multinomial ML models are adopted with 2,373 sample trip segments in 84 MTZs. The estimation results are listed in Table 6 column (a). Similar to the binary ML model, a high goodness-of-fit value (adjusted $\rho^2 = 0.885$) is obtained in the multinomial ML model considering BE effect, which shows these variables can well describe people's behavior. A control model without BE variables is estimated for comparative purposes. The estimates are shown in Table 6 column (b). A substantial decrease of goodness-of-fit (adjusted ρ^2 decrease from 0.885 to 0.813) can be found after discarding BE variables, which emphasizes the importance of BE factors on people's modal choice.

Referring to the estimates in Table 6 column (a), the coefficients of bus travel time, distance to MRT station, and EAI to bus stop provide the same implication as that in Table 5 column (a), indicating the robust effect of these variables on modal choice behavior. The ASCs of bus and LRT modes are not statistically significant, which means people have no inherent preference for these three modes when LRT is available.

As for LRT mode, the positive signs of distance to MRT station shows that, similar to "bus", people tend to use LRT when they are more distant from the MRT station. The distance coefficients of "LRT" (3.250) and "bus" (2.450) suggest that, keeping all other variables constant, increased distance from MRT station may encourage more people to use LRT than bus in the areas where the LRT is available. This is possibly due to the fact that LRT can take passengers directly to the MRT station, avoiding additional walking for interchange and the potential of encountering traffic congestion. According to the aggregate elasticity, if the distance to MRT station increases by 10% for each individual, the share of "walk" on average may decrease by 4.64% while "bus" and "LRT" may increase by 6.45% and 12.15%, respectively. Besides, the entropy is found to have negative effect on the utility of bus and LRT modes, which suggests that high levels of land-use mix will encourage more people to walk. This finding corresponds to the previous study (*14*).

Similar to the binary ML model, the standard deviations of physical BE variables are not significant in this model, while the coefficient of entropy for LRT mode is found to vary across the sample. The finding is consistent with the heterogeneity of impact of physical and socioeconomic BE in binary ML model.

Table 6. Modeling Estimation Results of the Multinomial ML Model

Variable		(a) Model with BE variables		(b) Model without BE variables	
		Coefficient	t-test	Coefficient	t-test
Walk					
Constant α	-	0	fixed	0	fixed
Travel time	Mean	–0.835	–4.06***	–1.260	–6.42***
	†SD	0.144	2.77**	0.235	4.28***
Bus					
Constant α	-	–3.860	–1.46	–7.290	–6.92***
Travel time	Mean	–1.850	–3.68***	–0.904	–5.57***
	SD	0.392	2.99***	0.154	2.70**
Distance to MRT station	†Mean	2.450	3.58***	-	-
	‡SD	1.430	0.16	-	-
Entropy (Non-MRT station area)	Mean	–15.40	–2.95***	-	-
	SD	0.439	0.34	-	-
EAI to bus stops (Origin)	Mean	3.020	2.76**	-	-
	SD	0.141	0.19	-	-
LRT					
Constant α	-	11.90	1.43	–7.790	–6.35***
Travel time	Mean	–3.230	–2.71**	–1.130	–6.11***
	SD	0.540	2.29**	0.008	0.11
Distance to MRT station	†Mean	3.250	2.69**	-	-
	†SD	0.032	0.23	-	-
Entropy (Non-MRT station area)	Mean	–44.40	–2.38**	-	-
	SD	3.600	1.73*	-	-
Statistics					
Observations		2373		2373	
Rho squared		0.891		0.816	
Adjusted Rho squared		0.885		0.813	

†Data scaled by 100.
‡Data scaled by 10,000.
***Significant at 99% level.
**Significant at 95% level.
*Significant at 90% level.

Conclusion and Discussion

The paper studies the impacts of BE on first- and last-mile travel mode choice based on the discrete choice model. We select Singapore as a case study. The dataset was 23,941 observations of first- and last-mile trips extracted from the HITS database. The BE factors were quantified using the four "D" variables proposed by Ewing and Cervero (*3*). In addition, sociodemographic variables and trip-specific variables were also taken into account in this work. To estimate the impact of BE and variation of taste, ML modeling frameworks are adopted. Since the availability of LRT may have a significant influence on travel behavior, two separate sample sets were used for performing a binary ML model (with walk and bus modes) and a multinomial ML model (with walk, bus and LRT modes), respectively. The models reveal the following findings. (a) BE—especially distance to MRT station, transportation infrastructures, land-use mix, and socioeconomic activities—significantly influences the first- and last-mile travel behaviors. (b) Those who live or work close to MRT stations and in an area with diverse socioeconomic activities and land-use mix may have stronger preferences toward walking for their first- and last-mile trips. (c) The impact of physical BE (i.e., distance, infrastructures) is relatively homogeneous across the sample, while the impact of socioeconomic-related BE (i.e., floor space density, entropy) varies.

Several policy implications are associated with the modeling results. From the point of view of urban design, increased probability of individuals choosing to walk would come from designing and building more compact communities with higher density of building floor space and proximity to MRT stations. Recent studies have revealed that walking can increase longevity and reduce the burden of important chronic conditions (*33–35*). This research provides meaningful suggestions for improving public health from the BE angle by promoting walking modes. The results also offer some suggestions for planning authorities to balance the demand for bus travel. For those who live further away from MRT stations, a bus system with high density of bus stops, better accessibility to the stops, and higher road network density, should be provided to meet their first- and last-mile travel demands. In addition, the results also offer a reference for the prospective implementation of autonomous vehicles (AV). The Land Transport Authority of Singapore launched the Singapore Autonomous Vehicle Initiative to explore the technology, applications, and solutions for AV in Singapore (*36*). Since AVs and buses are both motorized and shared travel modes for first- and last-mile trips, the areas with high first/last mile travel demand by bus may also imply high potential demand for AVs in the future. Therefore, the model results offer some suggestions for AV deployment and installation of infrastructure with consideration of BE to balance the use of different modes.

This study can be further improved from the following aspects. The first pertains to the assumption of ignoring self-selection bias. Due to the BTO policy, the common method with sociodemographic variables as the control may not be applicable here. Therefore, this assumption can only be further tested with more attitudinal data and a more advanced modeling approach, which is beyond the scope of the present study. Another path to improve this work relates to coping with multi-collinearity of data. The multi-collinearity problem results in the discarding of several variables such as the MRT station density. Future work can be done by applying the dimension reduction method (e.g., factor analysis, principal component analysis) to extract latent variables to illustrate the impact of BE better.

Acknowledgments

The research is supported by the National Research Foundation (NRF), Prime Minister's Office, Singapore, under the CREATE program, Singapore-MIT Alliance for Research and Technology (SMART) Centre, Future Urban Mobility (FM) Interdisciplinary Research Group. The authors thank Anson F. Stewart for his insightful comments and proofreading, and thank Daya Zhang for editing the figures. The first author also thanks the TOP OPEN program and Initiative Scientific Research Program of Tsinghua University, China, for financial support (20161080166).

Author Contributions

The authors confirm contribution to the paper as follows: study conception and design: B.M., Y.S., J.Z.; data collection: B.M., Y.S.; analysis and interpretation of results: B.M., Y.S.; draft manuscript preparation: B.M., Y.S., J.Z. All authors reviewed the results and approved the final version of the manuscript.

References

1. Roof, K. Public Health: Seattle and King County's Push for the BE. *Journal of Environmental Health*, Vol. 71, No. 1, 2008, p. 24.
2. Handy, S. L., M. G. Boarnet, R. Ewing, and R. E. Killingsworth. How the BE Affects Physical Activity: Views from Urban Planning. *American Journal of Preventive Medicine*, Vol. 23, No. 2, 2002, pp. 64–73.
3. Ewing, R., and R. Cervero. Travel and the BE: A Meta-Analysis. *Journal of the American Planning Association*, Vol. 76, No. 3, 2010, pp. 265–294.
4. Stead, D., and S. Marshall. The Relationships between Urban Form and Travel Patterns. An International Review and Evaluation. *European Journal of Transport and Infrastructure Research*, Vol. 1, No. 2, 2001, pp. 113–141.
5. Dieleman, F. M., M. Dijst, and G. Burghouwt. Urban Form and Travel Behaviour: Micro-Level Household Attributes and Residential Context. *Urban Studies*, Vol. 39, No. 3, 2002, pp. 507–527.
6. Frank, L., M. Bradley, S. Kavage, J. Chapman, and T. K. Lawton. Urban Form, Travel Time, and Cost Relationships with Tour Complexity and Mode Choice. *Transportation*, Vol. 35, No. 1, 2008, pp. 37–54.
7. Ewing, R., and R. Cervero. Travel and the BE: A Synthesis. *Transportation Research Record: Journal of the Transportation Research Board*, 2001. 1780: 87–114.
8. Cao, X. J., P. L. Mokhtarian, and S. L. Handy. The Relationship between the BE and Nonwork Travel: A Case Study of Northern California. *Transportation Research Part A: Policy and Practice*, Vol. 43, No. 5, 2009, pp. 548–559.
9. Cervero, R., A. Round, T. Goldman, and K.-L. Wu. *Rail Access Modes and Catchment Areas for the BART System*. University of California Transportation Center, Berkeley, CA, 1995.
10. Daniels, R., and C. Mulley. Explaining Walking Distance to Public Transport: The Dominance of Public Transport Supply. *Journal of Transport and Land Use*, Vol. 6, No. 2, 2013, pp. 5–20.
11. Park, S. *Defining, Measuring, and Evaluating Path Walkability, and Testing Its Impacts on Transit Users' Mode Choice and Walking Distance to the Station*. University of California, Berkeley, 2008.
12. Tilahun, N., P. V. Thakuriah, M. Li, and Y. Keita. Transit Use and the Work Commute: Analyzing the Role of Last Mile Issues. *Journal of Transport Geography*, Vol. 54, 2016, pp. 359–368.
13. Naharudin, N., M. S. S. Ahamad, and A. F. M. Sadullah. Assessing Criteria of the Pedestrian-Friendly First/Last Mile Transit Journey by Using Analytical Network Process (ANP) Group Judgment. *Proc., 6th International Conference on Social Sciences and Humanities*. Bangi, Selangor, Malaysia, April 2017, p. 203.
14. Ryan, S., and L. F. Frank. Pedestrian Environments and Transit Ridership. *Journal of Public Transportation*, Vol. 12, No. 1, 2009, p. 3.
15. Loutzenheiser, D. Pedestrian Access to Transit: Model of Walk Trips and Their Design and Urban Form

Determinants around Bay Area Rapid Transit Stations. *Transportation Research Record: Journal of the Transportation Research Board*, 1997. 1604: 40–49.
16. Jiang, Y., P. C. Zegras, and S. Mehndiratta. Walk the Line: Station Context, Corridor Type and Bus Rapid Transit Walk Access in Jinan, China. *Journal of Transport Geography*, Vol. 20, No. 1, 2012, pp. 1–14.
17. Shen, Y., H. Zhang, and J. Zhao. Integrating Shared Autonomous Vehicle in Public Transportation System: A Supply-Side Simulation of the First-Mile Service in Singapore. *Transportation Research Part A: Policy and Practice*, Vol. 113, 2018, pp. 125–136.
18. Singapore Land Transport Authority. *2012 Household Interview Travel Survey Background Information*, Singapore, 2012.
19. Zhu, Y., and J. Ferreira. Synthetic Population Generation at Disaggregated Spatial Scales for Land Use and Transportation Microsimulation. *Transportation Research Record: Journal of the Transportation Research Board*, 2014. 2429: 168–177.
20. Munshi, T. BE and Mode Choice Relationship for Commute Travel in the City of Rajkot, India. *Transportation Research Part D: Transport and Environment*, Vol. 44, 2016, pp. 239–253.
21. Cervero, R., and K. Kockelman. Travel Demand and the 3Ds: Density, Diversity, and Design. *Transportation Research Part D: Transport and Environment*, Vol. 2, No. 3, 1997, pp. 199–219.
22. Tracy, A. J., P. Su, A. W. Sadek, and Q. Wang. Assessing the Impact of the BE on Travel Behavior: A Case Study of Buffalo, New York. *Transportation*, Vol. 38, No. 4, 2011, pp. 663–678.
23. Minocha, I., P. Sriraj, P. Metaxatos, and P. Thakuriah. Analysis of Transit Quality of Service and Employment Accessibility for the Greater Chicago, Illinois, Region. *Transportation Research Record: Journal of the Transportation Research Board*, 2008. 2042: 20–29.
24. Zhao, P. The Impact of the BE on Individual Workers' Commuting Behavior in Beijing. *International Journal of Sustainable Transportation*, Vol. 7, No. 5, 2013, pp. 389–415.
25. Singapore Land Transport Authority. *Singapore Land Transport: Statistics In Brief 2012*, Singapore, 2012.
26. Train, K. E. *Discrete Choice Methods with Simulation*. 2nd ed. Cambridge University Press, 2009.
27. Badoe, D. A., and E. J. Miller. Transportation–Land-use Interaction: Empirical Findings in North America, and Their Implications for Modeling. *Transportation Research Part D: Transport and Environment*, Vol. 5, No. 4, 2000, pp. 235–263.
28. Schwanen, T., and P. L. Mokhtarian. What Affects Commute Mode Choice: Neighborhood Physical Structure or Preferences toward Neighborhoods? *Journal of Transport Geography*, Vol. 13, No. 1, 2005, pp. 83–99. https://doi.org/http://dx.doi.org/10.1016/j.jtrangeo.2004.11.001.
29. Housing & Development Board. *Build-To-Order (BTO)*. Singapore. http://www.hdb.gov.sg/cs/infoweb/residential/buying-a-flat/new/bto-sbf. Accessed July 30, 2017.
30. Ben-Akiva, M. E., and S. R. Lerman. *Discrete Choice Analysis: Theory and Application to Travel Demand*. Cambridge, Mass.: MIT Press, 1985.
31. Hensher, D. A., J. M. Rose, and W. H. Greene. *Applied Choice Analysis: A Primer*. Cambridge University Press, 2005.
32. Bierlaire, M. *PythonBiogeme: A Short Introduction*, Report TRANSP-OR 160706, Transport and Mobility Laboratory, School of Architecture, Civil and Environmental Engineering, Ecole Polytechnique Fédérale de Lausanne, 2016.
33. Pucher, J., R. Buehler, D. R. Bassett, and A. L. Dannenberg. Walking and Cycling to Health: A Comparative Analysis of City, State, and International Data. *American Journal of Public Health*, Vol. 100, No. 10, 2010, pp. 1986–1992.
34. Hamer, M., and Y. Chida. Active Commuting and Cardiovascular Risk: A Meta-Analytic Review. *Preventive Medicine*, Vol. 46, No. 1, 2008, pp. 9–13.
35. Celis-Morales, C. A., D. M. Lyall, P. Welsh, J. Anderson, L. Steell, Y. Guo, R. Maldonado, D. F. Mackay, J. P. Pell, and N. Sattar. Association between Active Commuting and Incident Cardiovascular Disease, Cancer, and Mortality: Prospective Cohort Study. *British Medical Journal*, Vol. 357, 2017, p. j1456.
36. Singapore Land Transport Authority. *Singapore Autonomous Vehicle Initiative*. https://www.lta.gov.sg/content/lta-web/en/roads-and-motoring/managing-traffic-and-congestion/intelligent-transport-systems/savi.html. Accessed July 25, 2017.

The Standing Committee on Transportation and Land Development (ADD30) peer-reviewed this paper (18-02288).

Sage Publications USA
sagepub.com

ISBN 978-1-5108-8176-1

CURRAN ASSOCIATES INC.
proceedings.com